Lecture Notes in Electrical Engineering

Volume 719

The book series *Lecture Notes in Electrical Engineering* (LNEE) publishes the latest developments in Electrical Engineering - quickly, informally and in high quality. While original research reported in proceedings and monographs has traditionally formed the core of LNEE, we also encourage authors to submit books devoted to supporting student education and professional training in the various fields and applications areas of electrical engineering. The series cover classical and emerging topics concerning:

- Communication Engineering, Information Theory and Networks
- Electronics Engineering and Microelectronics
- Signal, Image and Speech Processing
- Wireless and Mobile Communication
- Circuits and Systems
- Energy Systems, Power Electronics and Electrical Machines
- Electro-optical Engineering
- Instrumentation Engineering
- Avionics Engineering
- Control Systems
- Internet-of-Things and Cybersecurity
- Biomedical Devices, MEMS and NEMS

For general information about this book series, comments or suggestions, please contact leontina.dicecco@springer.com.

To submit a proposal or request further information, please contact the Publishing Editor in your country:

China

Jasmine Dou, Editor (jasmine.dou@springer.com)

India, Japan, Rest of Asia

Swati Meherishi, Editorial Director (Swati.Meherishi@springer.com)

Southeast Asia, Australia, New Zealand

Ramesh Nath Premnath, Editor (ramesh.premnath@springernature.com)

USA, Canada:

Michael Luby, Senior Editor (michael.luby@springer.com)

All other Countries:

Leontina Di Cecco, Senior Editor (leontina.dicecco@springer.com)

** **This series is indexed by EI Compendex and Scopus databases.** **

More information about this series at http://www.springer.com/series/7818

Sangeeta Singh · Rajeev Arya · M. P. Singh ·
Brijesh Iyer
Editors

CMOS Analog IC Design for 5G and Beyond

Springer

Editors
Sangeeta Singh
Department of Electronics
and Communication Engineering
National Institute of Technology Patna
Patna, Bihar, India

Rajeev Arya
Department of Electronics and
Communication Engineering
National Institute of Technology Patna
Patna, Bihar, India

M. P. Singh
Department of Computer
Science and Engineering
National Institute of Technology Patna
Patna, Bihar, India

Brijesh Iyer
Department of Electronics and
Telecommunication Engineering
Dr. Babasaheb Ambedkar
Technological University
Mangaon, Maharashtra, India

ISSN 1876-1100 ISSN 1876-1119 (electronic)
Lecture Notes in Electrical Engineering
ISBN 978-981-15-9864-7 ISBN 978-981-15-9865-4 (eBook)
https://doi.org/10.1007/978-981-15-9865-4

This Springer imprint is published by the registered company Springer Nature Singapore Pte Ltd.
The registered company address is: 152 Beach Road, #21-01/04 Gateway East, Singapore 189721, Singapore

Preface

Today is an era of exponential growth in wireless communication industries, specifically mobile data and the fifth-generation (5G) mobile network. This opens up new market opportunities in designing integrated circuits (ICs) required in this field. Till now, the CMOS technology has served this purpose well owing to its various inherent advantages viz., cost-effective mass production of digital ICs, scalable feature size that enables the integration of low power mm-Wave analog circuitry with the base-band digital signal processing units. This enables CMOS as the strong candidate for 5G and E-band mm-Wave circuitry designing; however, now the scaling of CMOS circuits is no longer advantageous. Now, this scaling trend is showing a bottle-neck in terms of maximum achievable f_{max} due to less efficient metal interconnect near the substrate, maximum achievable quality factor of these passive on-chip devices, and V_{DD} scaling also aggravates the designing trade-off as it affects the device linearity, device integration, and noise immunity. Moreover, at these scaled dimensions, the back end of line (BEOL) metal stack also gets scaled-down near substrate resulting in a significant effect of losses due to interconnects and parasitics. Furthermore, the design requirement of mm-Wave design and RF design is contradictory in terms of many FOM parameters. Thus, the research community has put a lot of effort to address these contradictory trade-off requirements for CMOS devices in the "Dark Silicon Era" for 5G and E-band communication circuitry. Recently, at 22 nm scaled nodes, FinFETs are outperforming conventional CMOS device structures. Based on this finding, researchers have explored various FinFET-based circuits for realizing 5G and E-band circuitry. Consequently, this book is focused on addressing the designs of FinFET-based analog ICs for 5G and E-band communication networks. In addition to this, it also incorporates some of the contemporary developments over different fields. It highlights the latest advances, problems, and challenges and presents the latest research results in the field of mm-Wave integrated circuits designing based on scientific literature and its practical realization. The traditional approaches are excluded in this book. We have covered various design guidelines to be taken care of while designing these circuits and detrimental scaling effects on the same. Moreover, gallium nitrides are also reported to show huge potentials for the power amplifier designing required in 5G communication network. Subsequently, to enhance the readability of this book, we have also included real-time problems in RFIC designing, case studies from experimental results, clearly demarking design guidelines for the 5G communication ICs designing. This book incorporates the most recent FinFET architecture for the analog IC designing and the scaling effects along with the GaN technology as well.

Patna, India Sangeeta Singh
Patna, India Rajeev Arya
Patna, India M. P. Singh
Mangaon, India Brijesh Iyer

Contents

About the Editors

Dr. Sangeeta Singh is an Assistant Professor, Department of Electronics and Communication Engineering, NIT Patna, India. She has received the Engineering degree in Electronics & Communication Engineering from SGSITS, (RGPV University, Bhopal) 2009, India, and the Master of Technology from Indian Institute of Information Technology and Management (IIITM), 2012, Gwalior, India. She received the Ph.D. degree in Communication Engineering from Indian Institute of Information Technology Design and Manufacturing (IIITDM Jabalpur) 2016, Jabalpur, India. She has been recognized as an eminent scholar in the field of Electronics & Computer Engineering. She has published 37 research papers in reputed national/international journals and conference. Along with this, she has actively participated in various technical courses workshops, seminar, etc., at the IITs and NITs. She reviewed various journals time to time, namely Nanotechnology IOP, Solid State Electronics, Elsevier, Electronics Letters IET, Micro Nano Letter, IET, Electronics Device Letter, IEEE, Journal of Nanoelectronics and Optoelectronics, American Scientific Publishers, Recent Patents on Engineering Advances in Science, Technology and Engineering Systems Journal and IEEE International Conference On Control, Automation, Robotics and Embedded System, CARE 2013. Her current research interests are comprehensive modeling of green transistors and charge based beyond CMOS devices, analytical and compact modeling of low power devices, modeling, simulation and characterization of emerging nano-electronics devices, development of low-cost medical monitoring, display and diagnostic biosensors using above devices and ultra-steep sub-threshold slope devices.

Dr. Rajeev Arya received the Engineering degree in Electronics & Communication Engineering from Government Engineering College, Ujjain, (RGPV University, Bhopal) 2008, India, and the Master of Technology in Electronics & Communication Engineering from Indian Institute of Technology (ISM), 2012, Dhanbad, India. He received the Ph.D. degree in Communication Engineering from Indian Institute of Technology (IIT Roorkee) 2016, Roorkee, India. He has received Ministry of Human Resource Development Scholarship

(MHRD India) during M.Tech. and Ph.D. He has been working as an Associate Professor in the Department of Electronics and Communication Engineering in CMR Engineering College Hyderabad. He is currently an Assistant Professor with the Department of Electronics & Communication Engineering at National Institute of Technology, Patna, India. His current research interests are in wireless communication, soft computing techniques, cognitive radio, signal processing, communication systems & circuits design. He has published many articles in international journals and conferences and received best paper award at ICCET-2019. He serves as Guest Editor of International Journal of Information Technology and Web Engineering (IJITWE) IGI Global Publishers USA, International Journal of Computational Systems Engineering INDERSCIENCE Publishers, UK, and Reviewer for IEEE Communication Letter, some journal of Springer Publisher, SCI/SEC-E, Scopus indexed journals and some conferences. He is a member of the IEEE, ISRD and the IAENG. He is an active reviewer in many reputed international journals.

Dr. M. P. Singh is an Associate Professor, Department of Computer Science and Engineering, NIT Patna, India. He has received the Master of Technology in Computer Science and Engineering from M.N.N.I.T., Allahabad, UP, India, Deemed University, Allahabad, India, in 2003. He received the Ph.D. degree in Computer Science and Engineering in 2008 from the same institute. He has supervised 9 Ph.D.s in the various domains of machine learning and wireless sensor networks, biomedical machine learning and energy-efficient clustered protocols. He has 8 journals and 70 conference research publications to his credit. He has also written 3 books and 4 book chapters, edited 3 books and reviewed 9 books. He has huge experience of project handling as he was PI of ISEA Project, Phase – II, DeitY, GoI, Nodal officer for Implementing NKN Project, DeitY, Government of India, for Establishing Virtual Classroom, DeitY and project under Ph.D. scheme of DeitY in 2014–2015, Government of India. His current research interests are fuzzy set, social media, machine learning and wireless sensor network. He has actively participated in various technical courses workshops, seminar, etc., at the IITs and NITs.

Dr. Brijesh Iyer received his Ph.D. degree in Electronics and Telecommunication Engineering from Indian Institute of Technology, Roorkee, India, in 2015. He is an Associate Professor in the University Department of E&TC Engineering at Dr. Babasaheb Ambedkar Technological University, Lonere (A State Technological University). He is a recipient of INAE research fellowship in the field of Engineering. He had 02 patents to his credit and authored over 40 research publications in peer-reviewed reputed journals and conference proceedings. He had authored 05 five books on curricula as well as cutting-edge technologies like sensor technology. He has served as the program committee member of various international conferences and reviewer for various international journals. His research interests include RF front-end design for 5G and beyond, IoT and biomedical image/signal processing.

Introduction to 5G Telecommunication Network

Mangal Das and Amitesh Kumar

Abstract In this chapter, the requirements and characteristics of the 5G network are discussed at an introductory level. 5G wireless networks operate at much higher frequencies (30–300 GHz) than their 4G predecessor networks. Higher data rates and density, lower latency (~ 1 ms) are some of the essential requirements of a 5G network. 5G communication network has three critical scenarios, such as Enhanced Mobile Broadband (eMBB), Ultra-Reliable Low Latency Communications (URLLC), Massive Machine Type Communications (MMTC), which constitute the bulk of requirement of a 5G network. A large number of stakeholders are critical issues with the 5G requirements. Lack of cooperation and regulation between stakeholders can lead to a situation of incoherent and paradoxical requirements for any system. There is no single technology that can fulfill the needs of all requirements of different stakeholders. Techniques like beamforming and software-defined networking can achieve such diverse requirements. Effective thermal management is needed to minimize the variations in the output of processing units and power amplifiers, which are used in a 5G network. Thermal management becomes critical at high frequencies as thermal conductivity and thermal coefficient of the dielectric constant of the substrate changes vary rapidly with temperature. The introduction of an active thermal management system reduces dispersion along interconnect due to variations in the dielectric constant. Overall 5G network poses a very complicated set of requirements to circuit and infrastructure designers.

Keywords 5G · Beamforming · MIMO · Path loss · Coupling loss

M. Das (✉)
Department of Electrical Engineering, Indian Institute of Technology Bombay, Powai, Mumbai 400076, India
e-mail: mangaldas111@iitb.ac.in

A. Kumar
Department of Electrical Engineering, Nextgen Adaptive Systems Laboratory (NASL), National Institute of Technology Patna, Patna, Bihar, India

S. Singh et al. (eds.), *CMOS Analog IC Design for 5G and Beyond*, Lecture Notes in Electrical Engineering 719, https://doi.org/10.1007/978-981-15-9865-4_1

1 Introduction

The 5th generation (5G) mobile network is an upcoming global standard of wireless communication after 1G, 2G, 3G, and 4G. The aim of 5G mobile network is to provide seamless virtual connectivity to machines, objects, and personal devices [1]. A 5G mobile network should consist of technologies such as mmWave propagation, smaller cells, massive multiple-input multiple-output (MIMO) networking in full-duplex mode. Such a complex system also necessitates software-defined networking (SDN), and beamforming.

Basic definitions [2] of the aforementioned terms are given below:

Millimeter wave: Millimeter waves cover frequencies of 30 GHz and 300 GHz, which are significantly higher than 4G frequency (< 6 GHz). Such high frequencies allow 5G networks to transmit data at higher speeds.

Small cells: Small cells are low-powered portable base stations that can be installed many small cells to form a dense, multifaceted infrastructure.

Massive MIMO: Massive MIMO infrastructure allows the transmitter and receiver to have multiple antennas, which maximizes the efficiency and speed. However, it may also introduce interference, which necessities the beamforming.

Beamforming: Beamforming is a vital part of the 5G cellular network. It increases the efficiency route of data delivery to individual users by steering narrow beams (in all three dimensions).

Full duplex: Full duplex communication can increase the speed of data transmission in 5G by nearly 2. However, this scheme may have a drawback of signal interference due to the use of the same channel for transmission and reception of data.

Software-defined networking (SDN): SDN will allow 5G network operators to create virtual subnetworks from a vast network. SDN decouples the network control and other functions, which make the controlling of network control easier.

2 5G and IoT Circuit Requirements

The 5th generation (5G) mobile network is an upcoming global standard of wireless communication after 1G, 2G, 3G, and 4G. Aim of Industrial internet of things (IoT) has some unique communication necessities such as high reliability, low latency, flexibility, and security, which can be fulfilled by the 5G mobile technology. This scenario makes 5G a popular contender for Industrial IoT applications. All of such high-level requirements, which are discussed in various sources, are standardized in ITU-R M.2083-0 [3]. ITU-R M.2083-0 standard defines three different crucial communication scenarios:

1. Enhanced Mobile Broadband (eMBB): 3GPP defined Enhanced Mobile Broadband (eMBB) as a 5G network use case. eMBB is a successor of 4G networks as it provides higher data rates for 360° video streaming, virtual reality, etc. Thus

it promises an enhanced consumer experience than present mobile broadband services [4]. However, the required performance can be ensured by adopting new modulation schemes with massive MIMO, SDN, etc. [5].

2. Ultra-Reliable Low Latency Communications (URLLC): Network reliability and safety during the transmissions were not a prominent topic in previous network generations. However, emerging telecommunication applications such as telesurgery, road safety, and industrial automation require massive effort in this area. To ensure such low latency and high safety, techniques like "network slicing" may some help [5].
3. Massive Machine-Type Communications (MMTC): In the future, many machines, sensors will be connected to the internet. In this case, all these systems (which have different architecture and data requirement) need to be connected. In this area, the so-called "vertical" industries could play a significant role in extending the telecommunications market [5].

However, all these service requirements are not related to IoT applications. Capabilities, which are mainly applicable for IoT, are given below [6, 7]:

- Slicing of network,
- Improving the Efficiency of resource management for IoT operations,
- priority-based policy control with optimized user plane
- Estimation of capability given network,
- Quality of service monitoring (especially for URLLC, vertical automation communication, and eV2X services),
- Positioning services for non-public low energy consuming networks,
- Cyber-physical control applications.

The stringent requirements of 5G present some difficult tasks to the designers of handsets and network infrastructures. High linearity and high-power efficiency power amplifiers (PAs) are required for 5G RF transmitters as such transmitters will use array MIMO antennas. The design of such low cost PA is complicated as it will require a very high level of integration with multiple phased antennas [8]. Nowadays, III–V semiconductor device technologies are still used to design PAs as they provide better frequency response, breakdown robustness [8]. Low-cost silicon-based laterally diffused MOSFETs are used for sub-3.5 GHz band design as it provides better output power (POUT) handling capacities. GaAs or GaN-based devices are used for microwave frequencies depending on the POUT requirements. For example, GaN-based devices provide an excellent power density of 3.6 W/mm at 86 GHz [9].

For applications, which require output power between 3 and 10 Watts and frequencies above 15 GHz, silicon-based power amplifiers have difficulties in competing with GaAs or GaN-based counterparts. GaN-based power amplifiers provide better alternative for the 5G technology with output power requirement of ~3–10 Watts [8].

3 5G Transmission Capacity Requirements

The 5G networks present some exciting tasks to the designers of mobile phones and communication infrastructures. 5G networks target the mm-Wave spectrum with eMBB, URLLC, and MMTC. The phased-array MIMO antennas with many radio frequency elements require an extraordinary integration with low cost [8]. 5G wireless networks operate at higher frequencies ranges compared with 4G and 3G. 5G systems will have 10–20 times higher data speed ($\leq$1 Gbps), about 1000 times denser, 10 times lower latency (~1 ms) compared to 4G LTE.

4 Microwave and MmW Transmission

Millimeter waves lie between the frequency range of 30–300 GHz of the electromagnetic spectrum. The wavelength range of millimeter waves varies from 10 mm (30 GHz) to 1 mm (300 GHz).

4.1 Requirements

ITU-R M.2083-0 [9] describes and analyzes the services and requirements such as transport capacity directly, network planning, networking affects throughput, site density (and MW/mmW link density), latency, slicing, agility (SDN etc.) respectively for 5G networks. International Mobile Telecommunications (IMT) have laid out the usage scenarios and required enhancements in the capabilities from IMT-Advanced to IMT-2020.

4.2 Spectrum

Millimeter-wave (mmWave) communication network has a frequency range from 30 to 300 GHz. Such a range allows nearly 250 GHz of bandwidth for communication. Such high frequencies of operation also give rise to problems like substantial path loss, atmospheric absorption, diffraction and blockage, and rain attenuation [10, 11]. However, millimeter waves make antenna element size smaller due to their small wavelengths. Smaller antenna elements allow the deployment in large numbers, which provides high spatial processing gains. In theory, high numbers of smaller antenna elements can be used to provide high processing gains (in 3-dimensional space) in order to compensate for isotropic path loss.

4.3 Transport Network Topology

5G network can use tree topology or meshed topology to function correctly. Tree topology is a hybrid hierarchical topology that interconnects star networks via bus networks. The Tree topology needs to recognize the explicit, active tail links and connect these links to one terminal, mobile site. A meshed topology also can be used by securing the fastest radio links. Meshed topology seems the most efficient way to ensure the secondary connection. To ensure good quality of service such a network may need to use SDN for network slicing for every path and service, and link protection capability with the ability to differentiate between different media [12].

4.4 Availability of Spectrum for Public Use

The availability of mmW spectrum for public usage depends on man factors such as political, defense-related issues, technological infrastructure, and regulatory factors. Technology is evolving at a rapid pace so that the full utilization of currently available spectrum (6–86 GHz) can be achieved. Moreover, currently higher frequency spectrums (W-band (100 GHz), D-band (150 GHz)) are under trials, which will enable higher data rates. The limited capacity of the current spectrum (6 GHz) is the primary reason to expand in the MW and mmW Spectrum.

The difference in the regulation and licensing (country-by-country) of the spectrum is a critical factor in the implementation of technological advancement into different countries [12]:

- mmW bands (especially above 23 GHz) are not available for public use in every country; this range of frequency is considered the "traditional" frequencies of the communication system.
- Present regulations are designed many decades ago for current lower frequency bands, based on the availability and capacity of communication infrastructure. However, modern features like adaptive modulation are very complicated and costly to implement under present regulation.
- New regulatory policies should address the issues related to the efficiency of the link spectrum by incorporating techniques like XPIC (Cross Polar Interference Cancellation), beamforming, and MIMO with high directivity antennas.

5 5G Networking Requirements

5G networks require high bandwidth, coverage, availability with low latency so that very high demands of faster communication can be accommodated. Table 1 gives a

Table 1 5G Network Requirements

Parameter	Requirement
Bandwidth	Min 1–10 Gbps
Latency	1 ms (transmission + reception route)
Number of connected devices	1–100 times of 4G
Availability	99.99%
Coverage	100%
Network energy usage	90% reduction from 4G
Battery life	~10 years (for low power handsets)

glimpse of some major requirements of a 5G network, which are gaining industry acceptance [13] (Table 1):

A large number of stakeholders are critical issues with the 5G requirements. Lack of cooperation and regulation between stakeholders can lead to a situation of incoherent and paradoxical requirements for any system. Each stakeholder will want their own needs to be met by the new 5G wireless system. There is no single technology that can fulfill the needs of different requirements of stakeholders [13]. Demand for data traffic is on the rise due to increased usage of images, email, web browsing, video, gaming, and live streaming, etc. by handset users. Data traffic requires faster handoffs between cells and data transmission/reception from tower to tower. A smaller cell may be used to increase the density of towers. However, it is an efficient, low-cost and technically complicated solution. It can become the foundation of a 5G network revolution [14].

6 mmWave Propagation Models

The diverse demands of large stockholders make designing a 5G wireless network very difficult. Designing of 5G networks requires a new kind of framework for channel models. Channel models of 5 G network should include more parameters for path loss (diffraction and interference), fading effect (large and small scale), and beamforming-enabled MIMO systems [15]. Different models have been discussed under different assumptions. The 3rd Generation Partnership Project model (3GPP) and NYUSIM model [15] are accepted by the International Telecommunication Union [15]. NYUSIM model has been developed after many years of field study in New York. 3GPP [15], ITU statistical models [15], and NYUSIM models use different parameters to accommodate the effect of beamforming by directional and steerable antennas, synchronized multipoint transmission. However, these models only simulate some of the real-world scenarios; the effectiveness of these models still needs improvement in many areas. For mmWave wireless systems, many dielectric and conducting obstacles of the real world will cause substantial path loss in comparison to free space. In order to improve current pathloss (PL) models, a detailed consideration of real-world parameters and in-field measurements has to be done. However, there are some empirical path loss models available for frequencies below 6 GHz. The next section discusses the alpha-beta-gamma (ABG) and close-in (CI) free space pathloss models for mmWave frequencies [16]. However, above the models (ABG and CI) are basic large-scale propagation path loss models that can be used in a particular scenario for all frequencies [16]. Small modifications in current 3GPP models give rise to CI model. This modification can be done by replacing a dynamic empirical constant in 3GPP model with a frequency-dependent constant.

6.1 Free Space Loss and Path Loss

Path loss for a radio wave can be defined as the reduction in power density as it propagates through the channel [15]:

$$PL[dB] = 10\log_{10}\frac{P_T}{P_R} \tag{1}$$

P_R is defined as a function of distance and wavelength (Friis' law) in a free space:

$$P_R(d,\lambda) = P_T G_T G_R \left(\frac{\lambda}{4\pi d_{3D}}\right)^2 \tag{2}$$

In a narrowband flat fading channel, small-scale fading can be given as the part of channel impulse response [15]:

$$h(t,\tau) = V + g(t,\tau) \tag{3}$$

where, V is a complex and deterministic component, which is defined for the line-of-sight path between the transmitter and receiver. If we assume the multiple received radio waves are wide-sense stationary uncorrelated scattering (t, τ) is a complex zero-mean Gaussian random variable with its envelope obeying the Rayleigh distribution.

The Alpha-Beta-Gamma (ABG) PL model [15] is given by:

$$PL^{ABG}(f,d)[dB] = 10\alpha\log_{10}\left(\frac{d_{3D}}{1m}\right) + \beta + 10\gamma\log_{10}\left(\frac{f}{1GHz}\right) + \chi_\sigma^{ABG} where\, d \geq 1m \tag{4}$$

where $PL^{ABG}(f,d)$ = mean path loss for a given f and d, χ_σ^{ABG} = zero-mean Gaussian random variable.

It is important to note that the ABG model has three model parameters (α, β, γ) for determining mean path loss and σ for determining shadowing. ABG model can be converted back to the 3GPP floating-intercept (AB) model for any single frequency by putting $\gamma = 0$ or 2 in Eq. 4.

The Close-in (CI) pathloss model for free space [15] is given below:

$$PL^{CI}(f,d)[dB] = 10n\log_{10}\left(\frac{d_{3D}}{d_o}\right) + FSPL(f,d_o)[dB] + \chi_\sigma^{CI}\,where\,d \geq d_o \tag{5}$$

where d_o = Close-in free-space reference distance, $\chi_\sigma^{CI}[dB]$ = zero-mean Gaussian random variable with a standard deviation σ in dB.

The CI model requires only two parameters n and χ_σ^{CI} whereas ABG model uses four parameters to calculate mean pathloss. The term "10n" in Eq. 5 is used to describe the path loss in terms of decades of distances beginning at d_o. This arrangement makes computation of power very easy for any distance in one's mind if d_o is set to 1 m. $FSPL(f,d_o)$ defines the free-space path loss (FSPL) if distance between transmitter and receiver is d_o and carrier frequency is f

$$FSPL(f,d_o)[dB] = 20\log_{10}\frac{4\pi f d_o \times 10^9}{c} \tag{6}$$

6.2 Outdoor-to-Indoor (O2I) Penetration Loss

The path loss in 3GPP channel model is a combination of three terms: outdoor path loss (PL_b), O2I penetration loss (PL_{tw}), and indoor path loss PL_{in} (depends on the depth into the building). The overall pathloss (for indoor users) in 3GPP model can be defined as [15]:

$$PL[dB] = PL_b + PL_{tw} + PL_{in} + N\left(0,\sigma_p^2\right) \tag{7}$$

σ_p^2 is the standard deviation for the penetration loss. The building penetration loss (Outdoor-to-Indoor) PL_{tw} through the external wall can be calculated by formula given below:

$$PL_{tw}[dB] = PL_{npi} - 10\log\sum_{i=1}^{N}\left(p_i \times 10^{\frac{-L}{10}}\right) \tag{8}$$

where PL_{npi} is an additional loss added to account for the external wall loss in case of nonperpendicular incidence. L is the penetration loss of any material i, p_i is the proportion of the ith material.

6.3 Coupling Loss Performance Analysis

In order to capture all sources of attenuation, analysis of coupling loss (CL) during wave propagation is significant for outdoor and indoor mobile stations. The coupling loss informs about the attenuation in the desired signal when signal travels from a serving cell to a mobile station [17]. CL is the same for both models (ABG and CI) as it does not depend on the power of transmitter. CL can be mathematically defined in following manner [17]:

$$CL = G_T + G_R - (PL(f_c, d) + L_{O2I}(f_c) + L_{OA}(f, d) - G_m) \tag{9}$$

Geometry of antenna array affects gain (G_m,) which represents the different smaller gain due to multipath transmission. In order to analyze CL, link loss is given below:

$$Link\ Loss = (PL(f_c, d) + L_{O2I}(f_c) + L_{OA}(f, d) - G_m) \tag{10}$$

Coupling loss in Eq. 9 decreases with increase (larger negative value) in the link loss (Eq. 10) [17]. Table 2 shows entities used in calculations of Path loss in ABG and CI model.

Table 2 Entities used in calculations of Path loss in ABG and CI model

Entities used in different Models	Symbols
Transmitted power (watt)	P_T
Received power (watt)	P_R
Transmitter antenna gain	G_T
Receiver antenna gain	G_R
Wavelength (m)	λ
Spatial distance in transmitter (m)	d_{3D}
Carrier frequency (GHz)	f
ABG Model Parameters (extracted from measured data)	α (depends on distance between transmitter and receiver), β (optimized offset value in dB for PL), γ (depends on carrier frequency), σ (standard deviation of large signal fluctuations in dB)
CI Model Parameter	n (path loss exponent) depends on distance between transmitter–receiver and carrier frequency)

7 Beamforming

Beamforming [18] is a technique in which highly directional antenna array elements can be used for spatial filtering during uplink and downlink in 5G networks. The phase angle between antenna elements can be tuned to create a constructive or destructive interference pattern on the desired location. Beamforming allows us to use the available spectrum in an efficient manner by changing the direction of antenna beam (even in milliseconds) in a massive MIMO array [19] Beamforming can be used for both electromagnetic waves and mechanical waves (sound). A beamformer is used in synchronization with antenna elements to control the phase and relative amplitude of the signal at each element. This arrangement actually makes wavefront patterns for constructive and destructive interference during communication. Beamforming has found applications in the areas of radar, sonar, seismology, wireless communications, acoustics. Beamforming can be divided into three types:

7.1 Baseband or Digital Beamforming

Digital beamforming sometime also called as precoding or baseband beamforming. In this scenario, the amplitude and phase of the antenna elements are preprogrammed to improve the cell capacity. This arrangement allows simultaneous transmission of data for multiple users with the help of effective management of resources (frequency/time). The same set of antenna elements can be used to form multiple beams (one per each user) simultaneously [18].

7.2 Analog Beamforming

Analog beamforming adjusts the amplitude and gain of the antenna array instantaneously, which allows partial compensation for high path loss at mmWave frequencies. However, analog beamforming only allows the formation of one beam for a set of antenna elements [18].

7.3 Hybrid Beamforming

Hybrid beamforming is a combination of above two beamforming schemes (analog and digital). One method is to use analog and digital beamforming for coarse and fine, respectively [18].

8 Geometry Metric Performance Analysis

Geometry metric (GM) performance analysis [17] provides the information about the quality of a desired signal in an small metropolitan situation for outdoor and indoor mobile stations. Moreover, this analysis includes the interference from other transmitting antennas and white Gaussian noise [17]. The GM can be defined as:

$$GM = \mathbb{E}\left(\frac{P_{RX,ser}}{N_o + \sum_{i \neq ser}^{N} P_{RX,i}}\right) \tag{11}$$

where $P_{RX,ser}$ and $P_{RX,i}$ are the received signal power from serving cell, and interference power from cell $i \neq ser$. N is the total number of sectors [17]. Geometry metric is the statistical analysis of the average signal-to-interference-plus-noise ratio in a given area.

9 Circuit Designer Perspective

9.1 Basic Requirements

5G wireless networks operate into mm-Wave frequencies (30–300 GHz, carrier frequency), which is higher than compared with 4G and 3G. The operating frequency range for 4G networks is 600 MHz to 6 GHz (carrier frequency) [20], along with available channel bandwidth of 20 MHz. 5G requires a large bandwidth to transmit very high date traffic, which mandates the design of high-speed, high gain power amplifiers. Such design requirements push mixed-signal design to its limits. Printed circuit boards (PCBs) will need to accommodate both at the same time. The channel bandwidth for the 5G network lies in the range from 100 MHz (<6 GHz) to 400 MHz (>6 GHz).

The bandwidth per channel is also an essential parameter that may require new ways to design PCB materials as it plays a crucial role in the performance of a system. Current integrated circuits can manage high data traffic and high frequencies. However, PCB materials will also need attention to ensure excellent performance for 5G systems [20]. 5G network requires a very high speed/high-frequency mixed-signal system. The material of component and PCB plays a vital role in preventing signal losses and ensuring signal integrity. Designers will need to use high-speed design rules and layout rules for optimizing operating conditions. Mixed-signal systems are prone to Electromagnetic interference (EMI). PCBs should be designed to prevent EMI between different sections (analog and digital) of the board, and design should meet electromagnetic compatibility requirements [20]. Such materials should have lower dielectric constant (~3); these new materials will able to replace polytetrafluoroethylene (PTFE) laminates for 5G wireless frequencies.

Effective thermal management is useful to minimize the variations in the output of processing units and power amplifiers. Thermal management becomes critical at high frequencies as thermal conductivity and thermal coefficient of the dielectric constant of the substrate changes vary rapidly with temperature. Typically, the dielectric constant of insulators follows inverse proportionality with temperature, which causes an increase in current as temperature increases. A thermal runway may gradually cause the loss of dielectric performance and an increase in power consumption [20].

To avoid heat-related issues, PCBs need to build on a highly thermally conductive substrate to dissipate heat away from active devices quickly. It will also stabilize the current in circuits, which in turn reduces variation in the output of processing units and power amplifiers at higher speed and frequencies. The introduction of an effective thermal management system will also reduce dispersion along interconnect due to variations in the dielectric constant. These dispersions along interconnect change RC time constants of parasitic capacitances, which in turn changes stretches digital pulses, change the propagation speed along an interconnect, and leads to signal reflections along a transmission line in extreme cases [20].

Author Contributions

This book chapter is written through contributions of all authors. All authors have given approval to the final version of the manuscript. Mangal Das and Amitesh Kumar are the main author and coauthor, respectively. The authors declare no competing financial interest.

Acknowledgements Amitesh Kumar is thankful to Department of Electrical Engineering, National Institute of Technology, Patna for providing resources to write this book chapter. Amitesh Kumar is thankful to Council of Scientific and Industrial Research to provide research fellowship to carry out research during his stay at Indian Institute of Technology, Indore.

References

1. Everything you need to know about 5G. https://www.qualcomm.com/invention/5g/what-is-5g. Accessed: 22 Apr 2020
2. Sd. Staff, 5G Technology promises faster connections, lower latency. https://www.sdxcentral.com/5g/definitions/5g-technology/. Accessed: 22 Apr 2020
3. IMT Vision—Framework and overall objectives of the future development of IMT for 2020 and beyond. https://www.itu.int/dms_pubrec/itu-r/rec/m/R-REC-M.2083-0-201509-I!!PDF-E.pdf. Accessed 23 Apr 2020
4. Kavanagh S (2020) What is enhanced Mobile Broadband (eMBB). https://5g.co.uk/guides/what-is-enhanced-mobile-broadband-embb/. Accessed 27 Apr 2020
5. ITU-T L.1310—Study on methods and metrics to evaluate energy efficiency for future 5G systems. http://handle.itu.int/11.1002/1000/13476. Accessed 23 Apr 2020
6. 3GPP TS 22.261 V17.1.0, 3rd generation partnership project; technical specification group services and system aspects; service requirements for the 5G system. http://www.3gpp.org/ftp//Specs/archive/22_series/22.261/22261-h10.zip. Accessed: 23 Apr 2020

7. Kozma D, Varga P, Soos G (2019) Supporting digital production, product lifecycle and supply chain management in industry 4.0 by the arrowhead framework - A survey. In: IEEE International conference on industrial informatics (INDIN), 2019, vol 2019, July, pp 126–131
8. Lie DYC, Mayeda JC, Li Y, Lopez J (2018) A review of 5G power amplifier design at cm-wave and mm-wave frequencies. Wirel Commun Mob Comput 2018:1–16
9. Niida Y, Kamada Y, Ohki T, Ozaki S, Makiyama K, Minoura Y, Okamoto N, Sato M, Joshin K, Watanabe K (2016) 3.6 W/mm high power density W-band InAlGaN/GaN HEMT MMIC power amplifier. In: PAWR 2016—Proceedings of the 2016 IEEE topical conference on power amplifiers for wireless and radio applications, 2016, pp 24–26
10. Lie DYC, Tsay J, Hall T, Nukala T, Lopez J (2016) High-efficiency silicon RF power amplifier design - Current status and future outlook. In: RFIT 2016–2016 IEEE International symposium on radio-frequency integration technology, 2016, pp 1–4
11. Bogale LBLTE, Wang X (2017) Chapter 9 - mmWave communication enabling techniques for 5G wireless systems: a link level perspective, a Paradigm for 5G," in mmWave Massive MIMO: A Paradigm for 5G, 2017, pp 195–225
12. Lombardi R (2018) Microwave and Millimetre-wave for 5G Transport, white paper, 1st edn. ISBN No. 979-10-92620-19-1, February 2018. https://www.etsi.org/images/files/ETSIWhitePapers/etsi_wp25_mwt_and_5g_FINAL.pdf. Accessed 23 Apr 2020
13. 5G Requirements: wireless technology needs. Electronics Notes. https://www.electronics-notes.com/articles/connectivity/5g-mobile-wireless-cellular/requirements.php. Accessed 28 Apr 2020
14. Nguyen T (2020) Small cell networks and the evolution of 5G (Part 1). https://www.qorvo.com/design-hub/blog/small-cell-networks-and-the-evolution-of-5g. Accessed 21 Apr 2020
15. Sun S, Rappaport TS, Shafi M, Tang P, Zhang J, Smith PJ (2018) Propagation models and performance evaluation for 5G Millimeter-Wave bands. IEEE Trans Veh Technol 67(9):8422–8439
16. Sun S, Rappaport TS, Rangan S, Thomas TA, Ghosh A, Kovacs IZ, Rodriguez I, Koymen O, Partyka A, Jarvelainen J (2016) Propagation path loss models for 5G urban micro-and macro-cellular scenarios. In: IEEE vehicular technology conference, 2016, vol 2016
17. Rupasinghe N, Kakishima Y, Guvenc I (2017) System-level performance of mmWave cellular networks for urban micro environments. arXiv:1708.03963v1
18. Passoja M (2020) 5G NR: Massive MIMO and Beamforming – What does it mean and how can I measure it in the field? https://www.rcrwireless.com/20180912/5g/5g-nr-massive-mimo-and-beamforming-what-does-it-mean-and-how-can-i-measure-it-in-the-field. Accessed 27 Apr 2020
19. Amy Nordrum ISS, Clark K (2020) 5G Bytes: beamforming explained. https://spectrum.ieee.org/video/telecom/wireless/5g-bytes-beamforming-explained. Accessed 25 Apr 2020
20. PCB materials and design requirements for 5G systems. https://resources.pcb.cadence.com/blog/2019-pcb-materials-and-design-requirements-for-5g-systems. Accessed 23 Apr 2020

Various Aspects of MOSFET Technology for 5G Communications

Pramila Jakhar, Amitesh Kumar, Mangal Das, and P. Rajagopalan

Abstract The idea of reducing the gate length of the metal-oxide semiconductor field-effect transistor (MOSFET) has been the leading stimulus for the growth of the integrated circuit industry. This chapter presents the current CMOS technology and its future advancements. The scaling concept and technology advancements in VLSI are discussed. Additionally, the transistor models and associated device modeling approaches are included. Herein, we discussed various aspects of CMOS integration into analog circuits for various architectures implementation and their desired improvements to work for futuristic 5G applications. The basic MOSFET device structure and modeling of MOSFETs for different signals and on different platforms considering various parasitics effects have been presented to make a way for the advanced research understanding of different works of modeling and proposed architectures in terms of simulation for their implementation for 5G. We expect that the future of CMOS with 5G will accompany different solutions arising with scaling issues and also provide various alternatives to make smooth incorporation of 5G into the wireless world.

Keywords MOSFET · 5G · CMOS scaling · MOSFET modeling

P. Jakhar (✉)
Department of Electrical & Electronics Engineering, BITS Pilani K K Birla Goa Campus, Goa 403726, India
e-mail: pramila@goa.bits-pilani.ac.in

A. Kumar (✉)
Department of Electrical Engineering, Nextgen Adaptive Systems Laboratory (NASL), National Institute of Technology Patna, Bihar, India
e-mail: amitesh.ee@nitp.ac.in

M. Das
Department of Electrical Engineering, Indian Institute of Technology Bombay, Powai, Mumbai 400076, India

P. Rajagopalan
College of Information Science & Electronic Engineering, Zhejiang University, Hangzhou 310027, India

S. Singh et al. (eds.), *CMOS Analog IC Design for 5G and Beyond*, Lecture Notes in Electrical Engineering 719, https://doi.org/10.1007/978-981-15-9865-4_2

1 Introduction

Theory of field-effect transistor (FET) reported by Lilienfeld in 1930 [1] and its practical implementation is shown on silicon/silicon-dioxide by Kahng and Atalla in 1960 [2] have steered the growth of FET, which became the key device of the semiconductor industry. The development in VLSI technology trails Moore's Law (1965). Recently, the sizes of the metal-oxide semiconductor (MOS) transistors viz. the respective channel length and width are scaled into deep nanoscale regime up to 10 nm. The hitech industry has been anticipating the advancements with the distinguished International Technology Roadmap for Semiconductors (ITRS) [3]. Technology developments marked in remarkable improvement in device density and performance, providing the basis for various data-intensive applications such as in autonomous driving, artificial intelligence, 5G communications, the Internet of things (IoT), and cyber-security.

2 Basics of CMOS Transistor and Scaling

Dr. Moore stated his empirical reflection that with the introduction of every new generation (node) of technology, the number of transistors doubles in the chip. The technological node is defined as the transistor's characteristic channel length, which changes the entire characteristics of the device. The node started with 180 nm channel length devices, which were successfully scaled down to as low as 5 nm. In a general trend of device scaling down, each generation of the device is reduced approximately 70% compared with previous generations. Simply, if we consider the square section, 70% reduction in channel length leads to 50% extra area, which leads to doubling of device density of fresher generation equated to prior generation node leading to a significant reduction in the cost of chip thus enabling a cost-effective solution in the reach of the common man. The conventional MOSFET device with a size of few millimeters in the 1970s could be scaled down to about a few nm in the recent year. In a MOSFET device, the electric field is varied under gate voltage, which can effectively tune the device in ON/OFF state as per the requirement. A MOSFET is a two-dimensional device where the current flows orthogonally to gate field. Due to such important characteristics, the MOSFET has been effectively utilized in the fields of memory devices, photovoltaic, sensors and computer processors. Moreover, the concept of scaling is not limited to the size of the transistor (which includes channel length, width, and thickness of the dielectric) but also includes shrinking the power supply to operate the device. The scaling of the device should be the completed in a judicious way in the horizontal, vertical and perpendicular directions. This feature of scaling ensures the scaled down voltage supply to the devices. Usually, in the literature, κ is referred as the scaling constant, which is usually 1.43 from one generation to another ($\kappa \approx 1.43 \approx 1/0.7$). Moreover, an important factor the doping concentration and charge densities should be appropriately adjusted for such scaling.

Thus, such scaling leads to condensed depletion regions between the source and drain. Robert H Dennard and his colleagues presented the first constant field scaling model which was constructed on simple electrostatic field theory [4]. The device was extremely small, which incorporated ion implantation to tune the required density. The model provided great deal on information, which are summarized as: (a) the circuit density is proportional to the square of scaling constant (κ^2), (ii) the processor speed is directly proportional to κ due to its relation with g_m/C (the capacitance of smaller dimensional devices is reduced by κ however the transconductance g_m does not affect with κ), (iii) power dissipation is inversely proportional to κ^2 [5]. Including the effects of all the parameters, the limit of scaling down is major concern. As we crawl down the lower dimensions, we reach a submicron level limit where the electric field in the channel twitches to alter significantly. The principle of broadly accepted gradual channel approximations, was first shown by Shockley [6]. This principle postulates that the electrical field in the MOSFET varies more slowly in the channel direction than in the gate direction. This principle is valid for all the device sizes except at the saturation region where it is invalid. Further, trimming the threshold voltage along with the total supplied power, the standby leakage current increases. Thus, increase in leakage currents constrains the reduction of power–supply voltage [7].

A vital aspect of gaining critically important improved performance in the transistor is to achieve a lower subthreshold slope (SS) (reciprocal of subthreshold swing). The SS denotes how abruptly the transistor current switches off lower the threshold value. Additionally, the carrier velocities get saturated, which is of more concern due to the nonscaling nature of the SS and the transistor off current I_{off}, which signify that the velocity saturation restricts the further speed enhancement with increasing the current. In the ideal situation, the subthreshold slope quickly turns off the device, which shrinks the value of leakage current whereas preserving the drive current value. Hence, a sharp slope could lower the threshold voltage (V_T) value, which in turn achieves a superior drive current ion while preserving the same I_{off}. These effects indicate that all the parameters must not be scaled by one factor. Sze et al. [8] and Dennard et al. [9] have solved this problem by giving the concept of general scaling. Sze gave the empirical relationship of scaling by $L_{min} = A.\left[X_j t_{ox}(W_x + W_d)^2\right]^{1/3}$.

Where L_{min} shows the minimum channel length, t_{ox} is oxide thickness, A denotes proportionality factor, X_j is junction depth, and $W_x + W_d$ is the sum of source and drain depletion depths in a 1D sudden junction formulation. This equation allowed a general miniaturization of device parameters than scaling of all parameters by the same factor.

The basic idea is that if the electric field intensity changes by a factor of ε (ε > 1, ε ≠ κ). Thus, the power supply voltage V_{DD} scaled by a factor of ε/κ. The electric field in the MOSFET is maintained constant by increasing the impurity doping concentration by ε. In case, these voltages cannot scale in any way, we have the fixed voltage scaling (V_1); in this condition, power does not scale; however, power density rises with κ^2 as each device continues to get smaller dimensions. The maximum integration density (the number of transistors per unit area) is measured either by the power dissipation density or by the area occupied by transistors, passive devices, and interconnections. According to the widespread scaling model, the active power density will be remarkably greater for the scaled devices due to the elevated electric field. Similarly, the performance significantly degenerates at larger V_T and also becomes more prone to changes in V_T with the decreased power–supply voltage. Subsequently, looking at the performance, as the power supply is downscaled, V_T should be reduced, and also the V_T variation affected by process tolerances should be reduced [5]. As we go toward each technology generation, the relative device density (i.e., the number of devices per chip) is reciprocal to the square of the relative lithography dimensions. These lithography directions define the minimum possible dimensions that can be realized on the die.

3 5G Circuits and Its Constraints

WIRELESS communication technology is continuously evolving and communication speeds are relentlessly increasing. The journey starts with 1G (first generation) in the 1990s, through 2G (second generation) of the communication method of GSM and CDMA, to 3G (third generation) of WCDMA, and now 4G (fourth generation) of LTE (Long-Term Evolution) and LTE-A (Long Term Evolution-Advanced). Talking about WRC-19, 5G millimeter wave (mmWave) spectra have been selected as the mmWave RF front-end arises to the turning point from concept to implementation. The millimeter-wave (mm-wave) band has materialized in response to the ever-mounting demand for a higher data rate in wireless communications. Mm-wave circuits and systems have been lately researched and developed to run into the necessity for a high data rate (>> giga-bit per second). With extensive bandwidth characteristics, the mmWave band, such as the 28 and 39 GHz, can be used to comprehend the Gbps data rate with high-order modulation schemes. From the execution point of view, there are many challenges. Such as there are challenging requirements for the VCO phase noise and tuning range. Moreover, for the portable wireless communication application, it is wished to realize VCO with low voltage and low power consumption [10]. To address these constraints of low voltage, wide tuning range and low phase noise issues of mmWave VCOs, and for making an allowance for the trade-off between the VCO phase noise and power consumption, researchers have been working a dual-core transformer feedback VCO which is the combination of the transformer feedback and VCO multicore techniques [10]. Similarly, there are other challenges. Like, in order to realize Gbps transmission rates, many researchers have voiced that the millimeter-wave should be used, but these waves do not penetrate solid materials quite well. They also tend to lose more energy than do lower frequencies over long distances, since they are readily absorbed or scattered by gasses, rain, and foliage. In order to compensate for high propagation attenuation, phased arrays are chosen to achieve array gain by beamforming. A phase shifter is a chief component to steer the beam of a phased-array antenna. Researchers propose a 4-bit digital phase shifter at 28 GHz, which is one of the candidates for the

5G frequency band, is designed and fabricated using a 1.0 V 1-poly 9-metal 65-nm CMOS process [11].

Besides, to overcome non-line-of-sight environments and limited communication distance, there is a need for the transceiver in adopted phased-array. Researchers are working on high linearity, image/LO rejection I/Q up-conversion mixer for 5G cellular communications. The mixer usages a symmetric layout method and a complementary derivative superposition (DS) technique with predistortion. The symmetric layout method improves image rejection ratio and LO leakage performances [12].

Besides, considering the general vision for fifth-generation (5G) wireless networks, which is a multitiered heterogeneous network having femtocells and picocells operating in the mm-wave band nestled within the existing network. At the range of millimeter wavelengths, wireless links are easily disordered when the line-of-sight (LoS) between the transmitter and the receiver is broken. Hence, to outspread a base station's coverage to "dark" locations that are "dark" in an office building or a city street, compact, low-cost two-way repeaters are required. Related with long-distance microwave backhaul links for long, there is renewed new interest in the design of microwave repeaters/relay nodes however this time it is for indoor links. A bidirectional amplifier circuit for fifth-generation (5G) communications systems is being worked out by researchers [13]. The circuit explained in this work [13], deploys two active quasi-circulators, one at the input and another at the output, and two amplifiers. The amplifiers are alike in design, with one fronting in the forward direction and the other the reverse direction [13]. We discuss more such works in the later sections where we consider various architectures, which would be useful to make way for the 5G world.

4 Metal Oxide Semiconductor Field Effect Transistor (MOSFET)

MOSFETs are the basic building block of MOS integrated circuits (IC). VLSI circuits utilizing MOS devices have become the primarily important in the semiconductor technology. Over the years, the intricacies of these MOS IC's have increased at a surprising rate [14]. In this section, the basic idea of MOSFET, its structure, and its operation has been presented.

4.1 *Basic Structure and Operation of MOSFETs*

4.1.1 MOSFET Structure

MOSFET is a four-terminal device with the terminals named as Gate, Drain, Source, and Body. The device is controlled by the electric field applied to the gate terminal. Based on the type of conducting channel whether it is n-channel or p-channel, MOS structures are classified as n-channel and p-channel MOS. Herein, the n-channel MOS (NMOS) transistor is discussed as both transistors are complementary. A p-type silicon substrate (substrate is also termed as body) is used in the fabrication of NMOS transistor. A low resistivity electrode is formed on the top center part of the device, which is separated from the body by an insulator. This low resistivity electrode is known as gate terminal. Usually, the gate material is made by using polysilicons with heavy doping of either n-type or p-type and silicon dioxide or simply oxide is utilized as an insulator. The source and the drain terminals are made by implanting donor impurities in two sides of the substrate [15]. The region between source and drain is called the channel having width-W and length-L, which plays a key role in determining the characteristics of a MOS transistor. Metal contacts are also provided to the source, drain, substrate, body (substrate and body are connected to the same terminal) region.

4.1.2 MOSFET Operation

For the MOS transistor, a current flow between the drain and source is controlled by the gate terminal voltage. When a sufficiently positive voltage V_{gs} is applied to the gate of NMOS, the positive charges are developed over the gate. These positive charges repel the holes of p-type substrate, which create negative charge acceptor ions and form the depletion region. Further increase in V_{gs}, at some potential level it will even make the surface attractive to electrons. Thus, an abundant of electrons are attracted toward the surface. As the surface of the p-type body normally has a large number of holes but with the application of higher V_{gs}, the surfaces now have a large number of electrons, hence this condition is called inversion.

With the application of positive potential across drain-to-source, electrons flow from the source and are drained by the drain terminal resulting a positive drain current I_d from drain-to-source. Under the proper biasing condition, two types of current can flow between the source and drain terminals of the MOSFET:

1. Diffusion current: Diffusion current flows when the applied gate voltage is less than the threshold voltage (i.e., $V_{GS} < V_T$).
2. Drift current: Drift current flows when the applied gate voltage is greater than that of V_T (i.e., $V_{GS} > V_T$).

Based on the situation of current flow between the drain and source, three operating regions exist [16–18]:

(1) Cutoff region: In the cutoff region, no current flows apart from subthreshold current, which becomes significant when the dimensions of the device are considerably smaller.
(2) Linear region: In the linear region, the linear current flows between the source and drain terminals.
(3) Saturation region: In this region, the current saturates because the conducting channel between the source and drain is pinched-off at the drain side with higher V_{DS} application.

4.2 *MOSFET Modeling*

The reduction of MOS transistor dimensions resulted in the complexity of MOS IC's over the past decade. With the increasing complexity, the design of such complex chips with high density transistors is difficult. This complexity in the design can be easily reduced with the help of simulation tools which can forecast the circuit behavior before the development of actual circuits. Although, the effectiveness of simulators for the circuit analysis mainly depends on the accuracy of the device models used in the simulator.

In this section, the MOSFET models based on basic semiconductor concepts have been discussed. Several models are developed that take into accounts all the observed physical effects in submicron devices utilized in current MOS VLSI technology. The development of these models is crucial when fabricating devices for MOS IC's. Herein, we will explore the output characteristics and the transfer characteristics of the MOSFET. In other words, we can find out both if we can formulate a mathematical equation of the form: $I_{DS} = f(V_{DS}, V_{GS})$

4.2.1 DC Model

Herein, the DC model of MOSFET circuits has been discussed. With zero gate voltage, no current flows from drain to source even with the application of a voltage V_{DS}. A very high resistance (of the order of 10^{12} Ω) path exists between drain and source. Consider next the situation when the source and the drain are connected to ground and a positive voltage is applied at the gate terminal. In this condition, as the source is grounded, the equivalent gate voltage appears between gate and source which is denoted by V_{GS}. The positive gate voltage forces the free holes (which are positively charged) away from the region of the substrate under the gate (the channel region) and thus creating a carrier-depletion region. Additionally, when the positive gate voltage is applied it pulls the electrons from the n^+ source and drain regions into the channel region. When a sufficient number of electrons are collected near the surface of the substrate under the gate, effectively an n region is created, between the source and drain regions. Now with the application of a voltage between drain and source, current flows through this induced n region. The value of V_{GS} at which a sufficient number of mobile electrons accumulate in the channel region to form a conducting channel is known as the threshold voltage V_t. The threshold voltage is positive for NMOS and it is in the range of typical value of 0.3 V–1.0 V. The gate and the channel region of the MOSFET create a parallel-plate capacitor, with the oxide layer used as the capacitor dielectric. When the positive gate voltage is applied a positive charge appears on the top plate of the capacitor, i.e., at the gate electrode. The equivalent negative charge appears on the bottom plate by the electrons in the induced channel. Thus an electric field is developed in the vertical direction. This electric field controls the charge in the channel, which relates to the channel conductivity and thus, the current will flow through the channel from source to drain when a voltage V_{DS} between drain and source is applied. This is the main reason that this device is named as "field-effect transistor" (FET). The voltage across the gate and channel formed, i.e., the voltage across the oxide, should be greater than V_t to create the channel. When $V_{DS} = 0$, the voltage at every point along the channel is zero, and the voltage across the oxide (i.e., between the gate and the points along the channel) is constant and equal to V_{GS}. The difference between V_{GS} and V_t is known as the effective voltage or the overdrive voltage and this quantity determines the charge in the channel.

The electron charge in the channel.

$$|Q| = C_{ox}(WL)(V_{GS} - V_t) \tag{1}$$

where C_{ox} is the oxide capacitance per unit area (F/m^2), W is the width, and L is the length of the channel.

The oxide capacitance C_{ox} can be calculated as:

$$C_{ox} = \frac{\epsilon_{ox}}{t_{ox}} \tag{2}$$

where ϵ_{ox} is the permittivity of the silicon dioxide, when the channel is induced, after application of a positive voltage V_{DS} between drain and source. When a small V_{DS} (i.e., 50 mV or so) is applied, a current i_D flows through the induced n channel. Hence, the current i_D flows in the channel from drain to source. Now, to calculate the value of i_D for small V_{DS}, it has been assumed that the voltage between the gate and each point through the channel remains approximately constant and equal to the voltage at the source end, V_{GS}. To calculate the current i_D according to the definition, it is the charge per unit channel length, which can be estimated from Eq. (3) as |Q| per unit channel length

$$\frac{|Q|}{L} = C_{ox} W V_{OV} \tag{3}$$

The voltage V_{DS} develops an electric field E across the channel length, which can be determined as:

$$|E| = \frac{V_{DS}}{L} \tag{4}$$

Due to this produced electric field, the channel electrons drift toward the drain with a velocity given by electron drift velocity v_n:

$$v_n = \mu_n E = \mu_n \frac{V_{DS}}{L} \tag{5}$$

where μ_n is the electrons mobility at the surface of the channel whose value depends on the fabrication process technology. The value of i_D can now be determined by multiplying the charge per unit channel length (6) by the electron drift velocity (Eq. 7),

$$i_D = \mu_n C_{ox} W L v_{OV} V_{DS} \tag{6}$$

Thus, for small V_{DS}, the channel acts as a linear resistance whose value is controlled by the overdrive voltage V_{OV}, Thus i_D can be rewritten in the form of Eq. 7

$$i_D = \mu_n C_{ox} W L (v_{GS} - V_t) v_{DS} \tag{7}$$

The conductance g_{DS} of the channel can be found from Eq. 8 or 8 as

$$g_{DS} = \mu_n C_{ox} \left(\frac{W}{L}\right)(v_{GS} - V_T) \tag{8}$$

$\mu_n C_{ox}$, depends on the process technology utilized for MOSFET fabrication. It is the product of the electron mobility μ_n, and the oxide capacitance C_{ox}. The channel conductance is proportional to each of μ_n and C_{ox} and thus to their product and called as the process transconductance parameter k_n' (A/V^2).

$$k_n^{'} = \mu_n C_{ox} \tag{9}$$

Second, the conductance g_{DS} as described in Eq. (8) depends on the transistor aspect ratio (W/L), which shows that g_{DS} is proportional to the channel width W and inversely proportional to the channel length L. The (W/L) ratio is determined by the device manufacturer. For a particular fabrication process, though, the minimum channel length, L_{min} is required. L_{min} that can be possible with for the fabrication process is used to characterize the process and is being constantly reduced as technology advances.

The product of the k_n' and W/L is called the MOSFET transconductance parameter k_n (A/V^2),

$$k_n = k_n^{'}\left(\frac{W}{L}\right) \tag{10a}$$

$$k_n = \mu_n C_{ox}\left(\frac{W}{L}\right) \tag{10b}$$

Moreover, the channel conductance g_{DS} depends on the overdrive voltage V_{OV}, which is a vital design parameter for circuits. Therefore, when V_{DS} is small, the MOSFET performs as a linear resistance r_{DS} whose value is controlled by the gate voltage V_{GS},

$$r_{DS} = \frac{1}{g_{DS}} \tag{11}$$

$$r_{DS} = \frac{1}{\mu_n C_{ox}\left(\frac{W}{L}\right)(v_{GS} - V_T)} \tag{12}$$

Thus, the MOSFET operates as a voltage-controlled resistance as demonstrated in Eq. 12.

The resistance is infinite for $V_{GS} \leq V_t$ and it decreases as V_{GS} is increased beyond V_t. Thus, for the MOSFET to conduct, a channel needs to be created. Finally, the current leaving the source terminal (i_S) is equal to the current entering the drain terminal (i_D), and the gate current $i_G = 0$.

When V_{DS} is increased, keeping V_{GS} at a constant value greater than V_t; i.e., assuming the MOSFET is operated at a constant overdrive voltage V_{OV}. Here, V_{DS} appears as a voltage drop across the length of the channel. As V_{DS} is increased, the channel becomes more tapered and its resistance increases correspondingly. Thus, i_D versus V_{DS} curve does not continue to follow the straight line and it gets saturated in this condition.

Thus, the drain current

$$i_D = \mu_n C_{ox}\left(\frac{W}{L}\right)(V_{GS} - V_T)V_{DS} - \frac{V_{DS}^2}{2} \tag{13}$$

This operation was based on the assumption that even though the channel became tapered, it still had a finite (nonzero) depth at the drain end. This condition is realized by keeping V_{DS} sufficiently small that the voltage between the gate and the drain, V_{GD}, exceeds V_t. To obtain this condition, V_{DS} should not be greater than V_{OV}, for as $V_{DS} = V_{OV}$, $V_{GD} = V_t$, and the channel depth at the drain end reduces to zero. The zero depth of the channel at the drain end gives rise to the term channel pinch-off. Increasing v_{DS} beyond this value (i.e. $V_{DS} > V_{OV}$) does not affect the channel shape and charge, and the current through the channel remains constant at the value reached for $V_{DS} = V_{OV}$,. The drain current thus saturates at a value found by substituting $V_{DS} = V_{OV}$,

$$i_D = \frac{1}{2}\mu_n C_{ox}\left(\frac{W}{L}\right)(V_{GS} - V_T)^2. \tag{14}$$

4.2.2 AC Model

MOSFET is generally used in linear amplifier circuits. For the mathematically analysis of the electronic circuit, a small-signal equivalent circuit of the MOSFET is necessary. The equivalent circuit involves capacitances and resistances leading to frequency effects on the MOSFET behavior. Initially, a small-signal equivalent circuit of MOSFET is introduced and then the involved physical factors limiting the frequency response of the MOSFET have been presented. One simplifying assumption, we will make in the equivalent circuit is that the source and substrate are both tied to ground potential. Two of the capacitances connected to the gate are inherent in the device. These capacitances are C_{gs} and C_{gd}, which represent the interaction between the gate and the channel charge near the source and drain terminals, respectively. The remaining two gate capacitances, C_{gsp}, and C_{gdp}, are parasitic or overlap capacitances. In real devices, the gate oxide will overlap the source and drain contacts because of tolerance or fabrication factors. As we will see, the drain overlap capacitance-C_{gdp}, in particular, will lower the frequency response of the device. The parameter C_{ds} is the drain-to-substrate pn junction capacitance, and r_s and r_d are the series resistances associated with the source and drain terminals. The small-signal channel current is controlled by the internal gate-to-source voltage through the transconductance. The voltage V_{gs} is the internal gate-to-source voltage that controls the channel current. The parameters C_{gsT} and C_{gdT} are the total gate-to-source and total gate-to-drain capacitances. This resistance is associated with the slope I_D versus V_{DS}.

In the ideal MOSFET biased in the saturation region, I_D is independent of V_{DS} so that r_{ds} would be infinite. In short-channel-length devices, in particular, r_{ds} is finite because of channel length modulation, which we will consider in the next chapter.

The source resistance r_s can have a significant effect on the transistor characteristics. The drain current is given by $I_d = g_m V_{gs}^{'}$ and the relation between V_{gs} and V_{gs}' can be found from

$$V_{gs} = V_{gs}^{'} + \left(g_m V_{gs}^{'}\right) r_s = (1 + g_m r_s) V_{gs}^{'} \tag{15}$$

The drain current can now be written as

$$I_d = \left(\frac{g_m}{1 + g_m r_s}\right) V_{gs} = g_m^{'} V_{gs} \tag{16}$$

The source resistance reduces the effective transconductance of transistor gain. The equivalent circuit of the p-channel MOSFET is the same as that of the n-channel except that all voltage polarities and current directions are reversed. The same capacitances and resistances that are in the n-channel model apply to the p-channel model.

4.2.3 Empirical Model

The empirical model is far more accurate than the previous generation. The JFET and MOSFET were improved drastically with dynamic memory allocation and modified nodal analysis for the smooth and accurate analysis. Several polynomial functions were included with adjustable time step control functions, which improved the performance of the platform. SPICE2G.6 was the last FORTRAN version of the second generation SPICE, which was released in 1983. The third and more advanced version of SPICE was written in C language in the year 1985. This included a graphical interface for viewing results. The newer version was equipped with polynomial capacitors, inductors, and voltage-controlled sources with reduced convergence problems. The version also includes MESFET, lossy transmission line, and nonideal switch models with improved geometry interfaces accommodating smaller geometrical features in the transistor. The version contained BSIM3, MOS9, and BSIM4 models with well-behaved smoothing functions. Here, in this section, we will discuss mainly the latest generation models of SPICE. SPICE contains several device models namely, Op-Amp model, diode model, FET, MOSFETs, BJT model, signal, filter and frequency models, etc. Here, we will discuss mainly the MOSFET model in SPICE.

4.2.4 SPICE Model

SPICE is a simulation platform devised at the University of California, Berkley to analyze the performance of integrated circuits. SPICE stands for Simulation Program with Integrated Circuit Emphasis was introduced in the year 1972 by L W Nagel and D O Paderson. In the year 1969, Prof Ron Rohrer created a simulation platform CANCER (Computer Analysis of Non-Linear Circuits Excluding Radiation) with L W Nagel who was then a master student. This platform could perform simple tasks like DC, AC, and transient analysis along with Shockley equations for diodes and Ebers-Moll equations for bipolar transistors. SPICE was an advanced version of CANCER which was released in the public domain. The first generation of SPICE included JFET and MOSFET devices, which were written in the FORTRAN language on a large mainframe computer. Moreover, the platform was based on the nodal analysis, which made it into the industry-standard simulation tool. However, its accuracy was poor and was inefficient to predict the real values. In the year 1975, the second generation of SPICE was launched that contains BSIM1, BSIM2, Level 28 (MBSIM) models. The platform included is complex. The power of computer simulation is more apparent when using these complex device models in the analysis and design of integrated circuits.

The simulators PSpice and Multisim are SPICE-based simulators. These simulators offer the user to choose various different MOSFET models. The SPICE model parameter values given by the semiconductor manufacturer include a parameter called LEVEL. The selected LEVEL characterizes the MOSFET model to be used by the simulator. However, these parameter values not always suggest the accuracy, nor the complexity of the selected MOSFET model. LEVEL = 1 denotes the simplest first-order model (called the Shichman-Hodges model). This model is based on the square-law MOSFET equations described in DC model of MOSFET.

For the simulation of a MOSFET circuit, the user requires to define the values of the model parameters and the dimensions of each MOSFET in the circuit being simulated. The most important parameters to be specified are channel length L and width W. The areas and perimeters of drain and source are important to specify for SPICE to model the body-junction capacitances (otherwise, zero capacitances are taken by default). The precise values of these geometrical parameters depend on the actual layout of the device. In order to calculate these dimensions, the assumption about a metal contact is to be made to each of the source and drain regions of the MOSFET.

Thus, using SPICE simulator, one can calculate the values of the parameters of the MOSFET small-signal model using the DC operating point (bias point), which further can be utilized by SPICE to perform the small-signal analysis (ac analysis).

4.2.5 MOSFET Parasitic Model

One of the major effects with scaling down of transistor is the presence of parasitic capacitance, i.e., poly-to-contact coupling along with the corner capacitance, which shows more significant effect with respect to intrinsic behavior.

With the scaling down of transistor dimensions, i.e., reducing the channel length, and transistor width diminishes the intrinsic capacitance effects and facilitates the improved circuit performance. However, the extrinsic parasitic capacitances viz. capacitance occurring between gate and contact, depending on design parameters like poly-to-contact distance, number of contacts, and contact spacing, which do not scale down in the similar fashion as with the further increase in transistor density, the distance between contacts and gate reduces, which decreases the contact resistance and increases the coupling capacitance. Moreover, for minimum transistor dimensions, the corner capacitance effects become significantly important.

(1) C_{ox}—It is the intrinsic capacitance consisting of the channel-gate capacitance, and the overlap capacitance. The channel-gate capacitance changes with the change in transistor width W and poly length (L_{poly}).
(2) C_{fringe}—C_{fringe} is defined as fringe capacitance. It is the extrinsic capacitances, which scales with transistor width W, number of contacts, contact spacing (S_{con}), poly-to-contact distances, and poly length (L_{poly}).
(3) C_{pc}—C_{pc} is the poly-to-contact capacitance, depending on the transistor dimensions and layout parameters.
(4) C_{cor}—It is the corner capacitance, which is not dependent on L_{poly}, as well as on layout parameters.

(5) C_{gb}—C_{gb} is the capacitance between gate and body, which depends on L_{poly}. This component is voltage dependent, which links to the channel, and a voltage-independent part.

5 MOSFET Frequency Response

The influence of parasitic capacitances of the MOS transistor can be included to form the small-signal model, which can be effective at high-frequency operation. The frequency response of MOSFET is significantly affected by two frequency limitation factors defined as channel transit time and capacitance charging time. If the charge carriers has the saturation drift velocity v_{sat}, then the transit time of charge carriers is $\tau_t = L/v_{sat}$ where L denotes the channel length. This transit time relates to the maximum frequency limit.

The second limiting factor is the gate or capacitance charging time. Neglecting r_s, r_d, r_{ds}, and C_{ds}, the subsequent equivalent small-signal circuit where R_L is a load resistance.

The input gate impedance in this equivalent circuit is now finite. In the circuit I_i is the input current and adding the currents at the gate node:

$$I_i = j\omega C_{gsT} V_{gs} + j\omega C_{gsT}\left(V_{gs} - V_d\right) \tag{17}$$

Similarly, adding the currents at the output drain node:

$$\frac{V_d}{R_L} + g_m V_{gs} + j\omega C_{gsT}\left(V_{gs} - V_d\right) = 0 \tag{18}$$

Using Eqs. (17) and (18) to eliminate the voltage variable V_d, the input current can be determined as:

$$I_i = j\omega\left[C_{gsT} + C_{gdT}\left(\frac{1 + g_m R_L}{1 + j\omega R_L C_{gdT}}\right)\right]V_{gs} \tag{19}$$

Usually, $\omega R_L C_{gdT} \ll 1$, therefore, we can neglect $\omega R_L C_{gdT}$ in the denominator of Eq. (20) which simplifies to

$$I_i = j\omega\left[C_{gsT} + C_{gdT}(1 + g_m R_L)\right]V_{gs} \tag{20}$$

The parameter C_M is the Miller capacitance, which is given by

$$C_M = C_{gdT}(1 + g_m R_L)$$

When the transistor is working in the saturation region, C_{gd} is zero, but C_{gdp} is a constant. The gain of the transistor is multiplied by this parasitic capacitance, which can be a significant feature in the input impedance. The cutoff frequency or transition frequency f_T is known as the frequency at which the magnitude of the current gain of the device is unity (or 0 dB), which can be explained as the condition when input current I_i is equal to the ideal load current I_d.

Where load current

$$I_d = g_m V_{gs} \tag{21}$$

The current gain magnitude can be calculated as

$$\left|\frac{I_d}{I_i}\right| = \frac{g_m V_{gs}}{\omega\left[C_{gsT} + C_M\right]V_{gs}} \tag{22}$$

To calculate the cut-off frequency, the magnitude of the current gain is made equal to unity, thus the calculated cut-off frequency is given as:

$$f_T = \frac{g_m}{2\pi\left(C_{gsT} + C_M\right)}$$
$$f_T = \frac{g_m}{2\pi C_G} \tag{23}$$

where C_G is equivalent gate capacitance.

The cutoff frequency f_T is proportional to the transconductance and inversely proportional to the equivalent gate capacitance. This capacitance is dependent on temperature, which gradually decreases as the temperature decreases from 300 to 100 K [19]. This reduction is partly because of the increased series capacitor of the gate and the carrier freeze-out of the substrate. At low temperatures, this depletion effect of the polysilicon gate is higher. Therefore, the polysilicon/SiO_2 interface broadens the depletion region resulting into an additional series capacitor with the gate oxide capacitor, which reduces the total gate capacitance.

The other important figure of merit of the transistor is f_{max}, which is the frequency at which the unilateral gain is unity. f_{max} is a better measure of a transistor's ability compared with f_T.

$$f_{max} = \frac{f_T}{2\sqrt{(R_G + R_S)\left(2\pi f_T C_{gd} + g_d\right)}} \tag{24}$$

Here, g_d denotes the drain conductance, R_G, and R_S signify the gate and source resistances, respectively.

The electrical conductivity of a semiconductor material is dependent on the product of free carrier density and mobility. The gate resistance increases as the free carrier density decreases, and the mobility reaches saturation when the temperature reduces below 50 K [19].

6 Effect of Scaling on MOSFET Integrated

Gordon E. Moore, proposed Moore's law way back in 1965, that the complete computing power goes to double nearly each 18–24 months required the dimension to scale. The scarcity of resources has been challenging the scaling of devices with the increasing use of available power and bandwidth. Various factors such as the advent of cloud, continuous interaction of big data, and immediate data are the new compulsions Ultra-low power, high-performance, high-speed devices requires abundant computing resources to tackle the huge amount of data being generated. Continued scaling of MOSFETs to deal with the device performance has always been the demand of electronics and computing world.

PPAC values for node-scaling as per moore has been a guiding force

(a) Power: > 35% less energy per switching at a specified performance, **(b) Performance**: > 15% more working frequency at the scaled voltage, **(c) Area**: > 35% less chip space footprint **(d) Cost**: < 30% added wafer cost—20% fewer die cost.

Many technological innovations have been guided by the above rule to work on new materials, new processes as well on modeling so as to cater to the need of ever-growing demand of the industry. The era of transistors before 130 nm node has been following Dennard scaling. Conventional MOSFET scaling was primary defined by Dennard in 1974. The various scaling rules of Dennard scaling have been depicted in Table 1. It depicts the variations in integrated circuit performance (scaling the circuit dimensions, voltages, and substrate doping). These changes are specified in terms of scaling factor K. It is being proposed that by 2024 P and N devices may be arranged on top of each other to give way to further reduction.

Considering Moore's Law for Packaging, it takes place with (a) dual in-line packages (DIPs) with < 16 I/Os leading to (b) peripheral quad flat packages (QFPs) from 64 to 304 I/Os. This further led to ceramic packages with increased I/Os up to about 1000 by the early 1990s.

There are many options to continue the proposed Moore's Law for packaging, such as:

1. Extend silicon package interconnections.

Table 1 Scaling results for circuit performance [20]

Device or circuit parameter	Scaling factor
Device dimension tox, L, W	$1/\kappa$
Doping concentration	No
Voltage V	$1/\kappa$
Current I	$1/\kappa$
Capacitance $\varepsilon A/t$	$1/\kappa$
Delay time per circuit VC/I	$1/\kappa$
Power dissipation per circuit VI	$1/\kappa^2$
Power density VI/A	1

2. Cultivate large-panel, low-C, and low-R inorganic GPE packaging.
3. Develop other panel embedding technologies without molding compounds and assembly.
4. Move on to optoelectronic interconnections from electronic interconnections

Similar to Moore's Law for ICs, cost is an important factor for Moore's Law for Packaging. There is gradual progress [20] in the relative cost reduction, as packaging progressed from one package family node to the next. Glass packaging, being pioneered and developed by Georgia Tech and its partners, is expected to reverse the cost trend.

7 Parasitic

Parasitics are considered to be undesirable quantities in physical circuit components. When we take extreme low-level geometric details of any physical layout, many unconsidered components such as routing tracks become significantly important. Considering the transmission line, not only the wire-length but also the shielding of wires is also an important contributor to changes in output. Therefore, the module and corresponding route planning is very vital part of system design.

Further considering the analog design, where transistor acting as an amplifier, the corresponding transceiver design would be important. The signals range in order of millivolts there. If we ignore the body effect, MOSFET (which can be considered as a three-terminal device) in a two-port model will have one of the terminals acting as both input as well as output.

The significant parasitics in any transistor is gate (Cg) capacitance, which can be divided into three different capacitances:

(1) Gate-to-source capacitance, Cgs;
(2) Gate-to-drain capacitance, Cgd;
(3) Gate-to-bulk capacitance, Cgb.

With the change in the excitation of the transistor, Cg will be fragmented. With the increase in frequency, parasitic quantities force transistors to drive in the suboptimal region [21, 22].

Other parasitic components can be:

1. **Parasitic device resistance:** It is projected that in present technologies, series resistance reduces the saturation current by 40% or more. Owing to the increase of current density, lower resistance with decreasing dimensions simultaneously is a huge challenge. Decreasing interconnect resistance arising due to scaling, and controlling source/drain series resistance are serious concerns.
2. **Parasitic device capacitance:** The parasitic capacitance between source/drain and gate terminal of device is anticipated to upsurge with scaling of technology. There is a prerequisite of low-κ spacer materials and air spacer to offer good consistency and etch selectivity for contact formation of S/D [23].

8 Transistor as a Switch

Considering various types of switches available, we have metal-oxide semiconductor (MOS) transistor-based switches to dominate as 98% of them for modern analog applications. Two such switches are: (i) the ground-side switch and (ii) the power-side switch [24].

For two transistor switches, transistor-based switch is OFF for an open circuit and that it is ON for a zero resistance (short circuit). NMOS (n-channel MOSFET) transistor includes two metal electrodes, drain (D) and source (S). The third terminal electrode (gate or G) with voltage Vin is there with the channel through an insulator. For the NMOS (n-channel MOSFET) transistor and switch is ON. Further, as +ve V_{in} is applied to the control terminal, channel develops more conducting [24].

Further to consider an example of an enhancement-mode n-channel MOSFET (NMOS transistor) is a four-terminal semiconductor device [24]. Whereas in discrete circuits, which use discrete transistor components, the body terminal is tied to the source terminal. Consequently, the NMOS transistor turns into the three-terminal MOSFET device (gate, drain, and source) like BJT. Though, it is no longer symmetrical [24].

9 Capacitors

Continuing from the Section Parasitics, as the frequency rises, parasitic elements will yield the transistor to work in the suboptimal regime, which yields to f_T and f_{max}. Transitional frequency is generally the smaller of the two quantities and is primarily influenced by *Cgs* and *Cgd* [21, 25]. MOSHFETs have the advantages of the (a) low gate leakage current like MOS structure and (b) high-density, high-mobility 2DEG channel AlGaN–GaN heterojunction. Authors [25] here recommend and exhibit a novel high-power RF switch based on a SiO_2 (dielectric layer)-AlGaN–GaN MOSH capacitor. The switch comprises of two MOSH capacitors coupled back-to-back (D-MOSH structure). When the DC bias is applied to the D-MOSH structure, the voltage drops are disbursed between two capacitors inversely proportional to shunting resistances.

10 Inductors

Considering inductors with 28 nm CMOS technology, a low-power receiver front end (RFE) with two-tap continuous time linear equalization (CTLE) using active inductor load for bandwidth extension can be designed [26]. The load of the amplifier can be resistive or passive inductive [26]. Furthermore, to make the CTLE area efficient, the passive inductor can be interchanged with an active inductor, a PMOS current-reuse transconductance. This alternative solution saves the chip area and also increases the bandwidth, with a minor penalty in the phase noise. The value of the inductance for the desired frequency (10 GHz) can be altered by the R_D transistor and the inductance range can be adjusted as per the theory [26], which depends on the g_m and R_D values of the active-inductor based load $M_3 - M_4 - R_D$ [26]. The zero and pole frequency tuning capability of the circuit are governed by the real and imaginary parts of the impedance as observed at the output node with tuning capability around 7.0 GHz–20 GHz.

We may consider the transfer function of the CTLE with active inductance and the corresponding response [26]. The CTLE has been designed and implemented in 28 nm low power CMOS technology [26]. Also in another work [13], we may observe the use of inductors in the bidirectional amplifier circuit for fifth-generation (5G) communications systems [13].

11 Transformers

Further considering transformers deployed with the architectures with CMOS displaying future applications for 5G, we have several examples. Here there is a work [10], where based on the transformer feedback topology, with the power-ground interconnect inductor and embedded decouple capacitor, the second harmonic real impedance is finally made at the VCO cross-coupled pair source terminal. For the recommended dual-core VCO topology, with the dual-core topology, the VCO layout can be realized in a symmetrical way, refining the VCO symmetry further [10].

Considering the VCO second harmonic impedance, the common-mode return path and layout of the transformer-based VCO (i) without the interconnect inductor and (ii) the embedded *DCAP*, cab be taken respectively. In contrast, with the *DCAP* and the coupled interconnect inductor, the VCO power and ground common-mode return path are reduced as well as well-controlled [10].

Fabricated in a 65 nm CMOS process, the dual core VCO core area is 0.24 $\times$ 0.5 mm^2 [10]. The output phase noise at 1 MHz offset and $1/f$ noise corner versus the operating frequency can be considered to further analyze the output phase noise and tuning range figure-of-merit (FoMT) [10]. In this work, based on the transformer feedback topology, with the power ground interconnect inductor and embedded decouple capacitor, the second harmonic real impedance is attained at the VCO cross-couple pair source terminal [10]. Further, considering another work [27], researchers have worked on design considerations and methodology for D-band transformer-based Class-AB gain-boosting power amplifiers (PAs) [27]. For SOI, the technology under consideration is 22 nm FD-SOI, flip-well (SLVT) nFET (n-channel field effect transistor) with a current-gain transit frequency (f_t)/maximum oscillation frequency (f_{max}) of 375 GHz/290 GHz [2] is preferred in this design [27]. The Z-parameter matrix (*ZT*) of a transformer T-section model with series resistance (R_p and R_s), has been considered. RF input and output pads are probed with 100 μm pitch waveguide Cascade Infinity probes [27]. The measurement results considered specify the stable performance of the four-stage PA design. The input power P_{in} at 140 GHz is swept in the first measurement. The results show that this PA achieved a P_{sat} of 8 dBm, a G_p of 22.5 dB, a peak PAE of 6.6%, and a peak DE of 7%. Meanwhile, a P_{sat} and a peak DE more than 6.3 dBm and 5% from 115 to 150 GHz are achieved [27]. In this work [27], the comprehensive design considerations and methodology for the silicon-based D-band Class-AB PA are offered. The implemented transistor's de-embedding approaches are designated. An efficient design flow for TMN, which cooptimizes the impedance transformation and passive efficiency, has been worked upon to hasten the design of multistage transformer-based PAs [27].

12 Transmission Lines

Parasitics has already been shown to be a significant concern for millimeter-wave circuits, which act as undesirable quantities in circuit components. Therefore, planning (cell/module/place/route) is extremely important in the layout. Considering various works on a transmission line, we consider here one of the works [21], where the Class-F Power Amplifier along with Quarter-Wavelength Transmission Line has been dealt. Here, adding further resonator circuits would upsurge the efficiency of the Class F amplifier, however, which is unfeasible. The resultant amplifier, also called as the Class-F∞ amplifier, is considered. The quarter-wavelength transmission line acts as an unlimited number of resonators and its efficiency can reach 100%. [21]. The drain as well as current waveforms for amplifier are considered.

Furthermore, as we consider transmission-line matching; which is essentially altered from lumped element matching. Here, Z_L is the impedance that is to be transformed and l is the transmission-line length, are considered [21]. Where passive filters have issues of high loss and low Q-factors, active filters primarily emphasize on loss compensation as well as Q-enhancement. A lossy circuit is considered by a transmission line, there this compensation can be realized with negative resistance, which is implemented as an active circuit. Here, the transmission-line resonator is installed to deliver equivalent values of inductance (Leq) and capacitance (Ceq) for filtering action, whereas p-type (RF pMOS) and n-type (RF nMOS or npn HBTs) lead to negative conductance ($-Gp$ and $-Gn$) in the organization of elements. The compensated resonant tank is considered, where the conductance of the resonant tank (Geq) is to be reimbursed for by the sum of $-Gp$ and $-Gn$ [21].

13 MOS as a Transconductor

Considering the realization of MOS as a Transconductor [28], we have an example of the work of an enhanced linear MOS transconductor that deploys both the adaptive biasing and source degeneration approaches [28]. The key aspect is to use as Integrated Analog Filters (IAF). IAFs may be accomplished using two methods: a) discrete-time and b) continuous-time deployments. There are three distinct procedures to execute integrated continuous-time filters: (a) Active-*RC*, (b) MOSFET-C, and (c) G-C [28].

Active *RC* arrangements deploy op-amps, resistors, and capacitors as passive frequency elements. They show excellent linearity, but generally necessitate large die space for resistors and/or capacitors. The corresponding tuning can be achieved only in a distinctive manner by deploying arrays of passive elements as well as the large value resistors, which may lead to significant thermal noise [28].

The simple differential MOS transconductor has been considered [28]. However, one of the simplest topologies to linearize the transfer characteristic of the MOS transconductor is the one with source degeneration deploying resistors [28]. However, the shortcoming of this configuration is the large resistor value required to realize a wide linear input range. Since in this case Gm = 1/R, the obtained transconductance is restricted to small values

By interchanging the degeneration resistors for two MOS transistors working in the triode region, the corresponding circuit [28] is acquired, where we consider for effortlessly matched transistors M_1–M_2, M_3–M_4, and overlook the body and channel length modulation effects as well as the transfer characteristic of this transconductor [28].

Further another MOS transconductor that combines the two linearization approaches have been presented here. Using adaptive biasing current sources, the circuit [28] is modified. The linear MOS transconductor has been fabricated by using the 0.35 m process from TSMC. The diagram of the entire circuit is considered [28]. The active load M_9–M_{10} is tuned by the common-mode feedback circuitry for regulating the output common-mode voltage to the desired value. The simulated and measured DC characteristics are considered [28].

Further, an improved linear MOS transconductor, conjoining two linearization methods has been offered. The circuit as described, here, has good tuning capability and it works as for both fully balanced and unbalanced input signals, with some linearity depreciation. In a practical implementation, the ultimate linearity performance is set by the matching precision of the MOSFETs. The circuit has been fabricated and experimental results agree with simulated linearity performance [28].

14 Scaling Effects

Continuing from the section "Effect of Scaling on MOSFET Integrated," and other scaling discussion in this chapter, we may consider other features of scaling. As shown in Table 2 [20], many complications arise from the fact that the cross-sectional area of conductors is decreased by K^2 while the length is decreased only by K. The IR drop in such a line is constant (with the decreased current levels), but is K times superior in comparison to the lower operating voltages. The response time of an unterminated transmission line is characteristically restricted by its time constant R_LC, which is unchanged by scaling. Current density in a scaled-down conductor

Table 2 Scaling results for interconnection lines [20]

Parameter	Scaling factor
Line resistance, $R_L = \rho L/Wt$	κ
Normalized voltage drop IR_L/V	κ
Line response time $R_L C$	1
Line current density I/A	κ

is increased by K, which leads to reliability concerns. In conventional MOSFET circuits, these conductivity problems are comparatively minor, but they become significant for linewidths of micron dimensions. The problems may be evaded in high-performance circuits by broadening the power buses and by evading the use of n + doped lines for signal propagation [20].

Also, as we move forward, we have various technologies that are being seen as "*Beyond CMOS for application-specific functions and architectures*" [23]. Considering the scaling technologies beyond 2030, MOSFET scaling will likely become ineffective and/or very costly. Entirely new, non-CMOS types of logic devices and may be even new circuit architectures are potential solutions. Such solutions ideally can be assimilated onto the Si-based platform to take advantage of the established processing infrastructure, as well as being able to include Si devices such as memories onto the same chip. Even early adoption of beyond CMOS technology and/or computing is likely to be adopted around 2024 frame by tunneling field-effect transistor (TFET) for ultra-low power applications and memristors for neuromorphic applications [23].

Also, CMOS logic and memory may together form the predominant majority of semiconductor device production. The types of memory considered in this chapter are DRAM and nonvolatile memory (NVM). The stress is on commodity, standalone chips since those chips tend to drive the memory technology. However, embedded memory chips are likely to follow the same trends as the commodity memory chips, commonly with some time lag. For both DRAM and NVM, detailed technology requirements and potential solutions are considered [23].

15 Conclusion

In this chapter, we considered various aspects of CMOS integration into analog circuits for various architecture implementation and their desired improvements to work for futuristic 5G applications. We considered CMOS from the basic level, in terms of deice physics and corresponding features, up to the advanced level of practical implementation. Modeling of CMOS for different signals and on different platforms considering various parasitics effects have been studied to make a way for the advanced research understanding of different works of modeling and proposed architectures in terms of simulation for their implementation for 5G. We expect that the future of CMOS with 5G will accompany different solutions arising with scaling issues and also provide various alternatives so as to make smooth incorporation of 5G into the wireless world.

Author Contributions
This book chapter is written through contributions of all authors. Amitesh Kumar and Pramila Jakhar are both first authors as they have worked together on this publication and contributed equally. All authors have given approval to the final version of the manuscript. The authors declare no competing financial interest.

Acknowledgements Authors are thankful to Department of Electrical Engineering, National Institute of Technology, Patna for providing resources to write this book chapter. Amitesh Kumar is thankful to Council of Scientific and Industrial Research to provide research fellowship to carry out research during his stay at Indian Institute of Technology, Indore.

References

1. Lilienfeld JE (1926) Method and apparatus for controlling electric currents, US Patent 1745175, pp 28–31
2. Lin Y-K (2019) Compact modeling of advanced CMOS and emerging devices for circuit simulation
3. Association SI (2016) International technology roadmap for semiconductors, 2015 results. Itrpv:1–37
4. Dennard R, Gaensslen F, Yu W-N, Rideout L, Bassous E, Le Blanc A (1974) Design of ion-implanted MOSFET' S with very small dimensions. IEEE J Solid State Circuits 9(5):257–268
5. Davari B, Dennard RH, Shahidi GG (1998) CMOS scaling for high performance and low power—The next ten years. Low-Power C Des 83(4):47–58
6. Shockley W (1952) A unipolar 'field-effect' transistor*. Proc IRE 40(11):1365–1376
7. Angelov GV, Nikolov DN, Hristov MH (2019) Technology and modeling of nonclassical transistor devices. J Electr Comput Eng 2019
8. Brews JR, Fichtner W, Nicollian EH, Sze SM (1979) Generalized guide for Mosfet miniaturization. Adv Chem Ser 2(3):10–13
9. Baccarani G, Wordeman MR, Dennard RH (1984) Generalized scaling theory and its application. IEEE Trans Electron Devices 41(4):1283–1290
10. Fu Y, Li L, Wang D, Wang X (2020) A -193.6 dBc/Hz FoMT 28.6-to-36.2 GHz Dualcore CMOS VCO for 5G Applications. IEEE Access 8:1–1
11. Shin GS, Kim JS, Oh HM, Choi S, Byeon CW, Son JH, Lee JH, Kim CY (2016) Low insertion loss, compact 4-bit phase shifter in 65 nm CMOS for 5G applications. IEEE Microw Wirel Components Lett 26(1):37–39
12. Byeon CW, Lee JH, Lee DY, Kim MR, Son JH (2015) A high linearity, image/LO-rejection I/Q up-conversion mixer for 5G cellular communications. Eur Microw Week 2015:345–348
13. Wang L, Saavedra CE (2018) 28–31 GHz Bi-directional amplifier for 5G wireless repeaters. In: IEEE international symposium on personal indoor mobile radio communication PIMRC, vol 2017, October, pp 1–4
14. Arora N MOSFET models for VLSI circuit simulation, 1st edn. Springer, Wien New York
15. Sood H, Srivastava VM, Singh G (2018) Advanced MOSFET technologies for next generation communication systems—perspective and challenges: a review. J Eng Sci Technol Rev 11(3):180–195
16. Sedra KCS, Adel S (2018) Microelectronic circuits, 7th edn. Oxford University Press, p 1395

17. Neamen DA (2006) Semiconductor physics and devices basic principles, vol 9, no 5
18. Streetman BG, Banerjee SK (2014) Solid state electronic devices, 7th edn
19. Hong S, Choi G, Baek R-H, Hee-Sang K, Jung S-W, Jeong Y-H (2008) Low-temperature performance of nanoscale MOSFET for deep-space RF applications. IEEE Electron Device Lett 29(7):775–777
20. Dennard RH, Gaensslen FH, Yi H. Design of lon-implanted MOSFET' s with very small physical dimensions. IEEE J Solid-State Circuits SC-9:256–268
21. Božanić M, Sinha S (2020) Millimeter-wave integrated circuits: methodologies for research, design and innovation, vol 658. Springer
22. Lindelöw F, Garigapati1 NS, Lasse Södergren1 MB, Erik L (2017) III-V nanowire MOSFETs with novel self-limiting -ridge spacers for RF applications To. Semicond Sci Technol:0–7
23. IRDS (2016) International Roadmap for Devices and Systems 2017 Edition more Moore. IEEE Adv Technol. Hum 3027:1–36
24. Makarov SN, Ludwig R, Bitar SJ (2019) Practical electrical engineering
25. Simin G, Koudymov A, Yang ZJ, Adivarahan V, Yang J, Khan MA (2005) High-power RF switching using III-Nitride metal-oxide-semiconductor heterojunction capacitors. IEEE Electron Device Lett 26(2):56–58
26. Thulasiraman D, Gaggatur JS (2020) A tunable, power efficient active inductor-based 20 Gb/s CTLE in SerDes for 5G applications. Microelectron J 95:104657
27. Tang X, Nguyen J, Medra A, Khalaf K, Visweswaran A, Debaillie B, Wambacq P (2020) Design of D-Band transformer-based gain-boosting class-AB power amplifiers in silicon technologies. IEEE Trans Circuits Syst I Regul Pap:1–12
28. Kuo KC, Leuciuc A (2001) A linear MOS transconductor using source degeneration and adaptive biasing. IEEE Trans Circuits Syst II Analog Digit Signal Process 48(10): 937–943

CMOS Scaled Architecture and Circuit Choices for 5G

C. Shalu and Amitesh Kumar

Abstract In system and circuit designs for 5G, new challenges are encountered to build an efficient, high-performance phased array transceivers as 5G covers a huge amount of available spectrum in the millimeter wave bands. Basically, this chapter focuses on the CMOS circuit and system trades for large millimeter wave power consumers such as power amplifier, local oscillator generation and distribution in the framework of beamforming system design and mobile link requirements. An appropriate millimeter wave transceivers for 5G and their application for E-band 71–76 GHz circuit blocks in Intel 22FFL FinFET process have been discussed in detail.

1 Introduction

In the last several decades, a prospect for fifth-generation (5G) cellular communications to simplify large throughput with spectrally effective systems has driven the industry further in the sub GHz bands [1–5]. The "millimeter-wave (mmWave) bands," which has a frequency range of 30 GHz–300 GHz, cover the tens of GHz of the underutilized spectrum. This signifies an open edge to talk about the huge development in the wireless data traffic [5–9].

The scheming of transceivers at mmWave for the cellular system increases several issues in construction and circuit concerning the phased arrays and the extraordinary carrier frequency that does not come across in the preceding cellular systems [7–10]. Upon increment of carrier frequencies to influence the superior bandwidth of channel and greater free-space loss as anticipated by the Friis equation, it can be contradicted by enhancing antenna physical area for wavelength to preserve a continuous effective

C. Shalu (✉)
National Centre for Flexible Electronics, Indian Institute of Technology Kanpur, Kanpur, India
e-mail: shalu268@gmail.com; shalu@iitk.ac.in

A. Kumar
Department of Electrical Engineering, Nextgen Adaptive System Laboratory (NASL), National Institute of Technology, Patna, Bihar, India

S. Singh et al. (eds.), *CMOS Analog IC Design for 5G and Beyond*, Lecture Notes in Electrical Engineering 719, https://doi.org/10.1007/978-981-15-9865-4_3

antenna aperture [2, 10–12]. The subsequent high gain antennas are integrally directional and some form of beam steering is required to preserve the spatial coverage [13]. A natural choice for this purpose is phased-array systems as they simplify beamforming by an electronically steerable antenna array [12–15]. Phased arrays permit the reduced performance than the discrete array elements, such as less output power and higher noise figure, in comparison to whole array performance as its implementation is well aligned with the Si [1, 14]. Additionally, they emerge as a decent application for the integration in large scale to work sophisticated digital processing and calibration algorithms [14–16]. The array elements across multiple dies and packages need signal path networks at the carrier frequency or IF (Intermediate frequency) [17]. They may comprise the generation and distribution of a coherent local oscillator (LO) signal through the array. LO distribution consumes the largest power inside the system, due to its phase noise necessities and the integrally low efficiency in the distribution of the mmWave signal. a programmable phase shifter is required at every element in the beamforming, before else after the mixer or maybe inside the LO [16–18].

Additionally, PA (power amplifier) can also be tremendously power starved as the efficiency in the millimeter-wave frequencies can be inadequate because of the operating vicinity of peak frequency for the power gain, or modern silicon technology's f_{max} [17–19]. In this chapter, the prominent requirements and successive architecture choices challenged by the designers who are aiming at the design of millimeter Wave transceivers in CMOS scale are highlighted. The selections and compromises of a phased array for implementation in CMOS scale have been discussed. An advanced survey regarding the above-discussed points has been presented.

2 Basic Introduction and Definitions

2.1 Beamforming

In beamforming, an array of antennas are to be steered to transmit radio signals in a particular direction [2]. The antenna arrays having beamforming, estimate the direction of a strong beam of signal to be transmitted in any particular direction rather than broadcasting it into all directions. This technique is very much suitable in 5G where we need improved coverage and uninterrupted signal strength. Also, it helps in improvement in high data rates by optimizing the transmission and reception efficiency by using beamforming [2]. In this method, every antenna element in the antenna array is fed individually with the data signal, which is to be transmitted. The amplitude and the phase of each signal are then added constructively and destructively by the concentration of the energy in a narrow lobe or beam.

2.2 *Millimeter Wave (MmWave)*

Millimeter waves are also called as extremely high frequency (EHF) waves. These are a band of radio frequencies and are suitable for 5G networks [1, 2]. In previous mobile devices, frequencies less than 5 GHz are used. Millimeter-wave technology uses frequencies lie between 30 GHz and 300 GHz for data transmission. The name "Millimeter Waves" is because of its wavelengths between 1 mm and 10 mm, whereas currently available smartphones and other devices use radio waves, which are having wavelengths nearly several dozen centimeters [2].

2.3 *Phased Array*

In an antenna array, the radio signal current from the transmitter is fed to the individual elements or antennas having the exact phase relationship so that the radio waves from the single antennas add together to rise the radiation in a required direction while stopping to suppress radiation in undesired directions [2–5]. In a phased array, antenna transmitter power is supplied to the antennas via phase shifters, which are controlled by a computer system. The phase can be altered electronically hence the beam of radio waves can be steered into different directions. Meanwhile, the array must be comprised of many small antennas to acquire high gain hence; phased arrays are mainly useful to the high-frequency end of the radio spectrum, i.e., in the UHF and microwave bands where antenna elements are relatively small [2].

3 Circuit Choices and Architectures of Fifth 5G

In the last few decades, the prospects for 5G communication have driven the industry afar the sub GHz bands, to enable extraordinary throughput with spectrally effective systems. The "millimeter-wave bands," which are in the frequency range from 30 GHz to 300 GHz, contain tens of GHz frequency bands underutilized and further epitomize an open leading edge to report the huge development of the wireless data traffic. The design of millimeter wave transceivers for the mobile apparatus advances several architectural- and circuit-related concerns such as phase arrays and an extraordinary carrier frequency, which does not come across in the preceding cellular systems. At larger carrier frequencies than channel bandwidth higher free-space loss can be counteracted by enhancing the antenna physical area proportionate to the wavelength to maintain the effective antenna aperture constant, as predicted by Friis' equation [1, 2].

The subsequent high gain antennas are intrinsically directional and they need a certain arrangement of beam routing to preserve the spatial coverage. For that purpose, phased array systems are an accepted alternative as they enable the formation of the beam via an electronically driven antenna array. Additionally, phase arrays line up quite fine with the silicon implementation, which permits reduced performance of a discrete array element such as less power output and more noise figure, in comparison to the in total performance of an array [1, 5]. Additionally, these are a decent alternative for integration in large scale to work with sophisticated digital processing and calibration algorithms.

Array elements spanning over multiple dies/packages necessitate the signal path contacts at the carrier frequency or intermediate frequency. They may include generation and distribution of a coherent LO (local oscillator) signal through an array. The distribution of LO consumes large power in system because of phase noise necessities and the integrally less efficiency as well in the distribution of millimeter-wave signals. In the beamforming process, programmed phase shifter, for every element, is required to be employed before or after the mixer else in the LO itself.

The implementation and phase shifter position choice depend on the technology and system architecture. Finally, PA can be very power ravenous as efficiency at millimeter-wave frequencies is inadequate. The main aim behind this chapter is to call attention to the outstanding necessities and succeeding architecture solutions as challenged via millimeter-wave transceiver designers in the scaled CMOS architecture. We propose the selections and compromises of phased array implementation in CMOS scale. Furthermore, a study of the futuristic for the facts as debated above is presented.

4 Link-Level Requirements

In the last few years, the attention in millimeter-wave spectrum for the subsequent generation technology had covered the frequency bands in the range of 24 GHz–100 GHz. The fifth-generation NR (New Radio) millimeter-wave bands in 5G are projected to counterpart lesser frequency bands generally less than 6 GHz. NR mm bands are presently quantified in 3GPP Release 15 to comprise the spectrum at 24.25 GHz to 29.5 GHz and 37 GHz to 43.5 GHz. The NR bands have been identified as TDD with the channel bandwidth of 400 MHz. Further, additional bands take noteworthy attention in industrial applications, which are further under an active assessment for propagation modeling and global standardization. These bands freshly unlocked the unlicensed spectrum by the FCC at 64–71 GHz, and licensed spectrum in the E-band at 71–76 GHz and 81–86 GHz. The work presented herein mainly emphasizes the anticipated applications in the E-band 71–76 GHz, that is responsible for the boundless amount of spectrum and welfares for dense placement both.

Typically, the uses of fifth-generation millimeter wave are fairly small range and large bandwidth communications for UE (user equipment) vehicle to network, and base station backhaul. On the other hand, the system range aims over several hundreds of meter for the bands of 28 GHz and 39 GHz. The access range of 60 GHz and 70 GHz frequency can be directed at several tens of meters. To set up the transceiver necessities, assumptions, and a simple baseline link budget for an E-band 71–76 GHz

Table 1 Summary of link assumptions [1]

Parameter	Value
Frequency	73.5 GHz
Bandwidth	2 GHz
Patch gain (directivity + efficiency)	3 dBi
Output power per patch (including front-end loss)	3 dBm
Rx noise figure (including front-end loss	6 dB
Required SNR (16-QAM, 7/8 coding)	15 dB
Friis path loss per meter	99.3 dB
Total system margin	7 dB

system using single-carrier 16-QAM modulation is listed in Table 1. For urban micro-cell (UMi) scenarios, path losses are presented for the free space, LOS, and NLOS (non-LOS) 6–100 GHz models, as detailed in 3GPP TR38.901 [1, 14].

The conventions for output power, noise figure, and array size of the transceiver are centered on the preceding art [1, 3–7]. The appropriate modulation schemes for millimeter-wave systems comprised of single-carrier and multi-carrier waveforms, with both having advantages and disadvantages that depend over application. From the designing prospects of the transceiver, single-carrier waveform is more striking compared with their multicarrier counterpart such as OFDM, because it has considerably low PAPR (peak to average power ratio) and acceptance range of LO phase noise [8]. Suppose a persistent saturated power for the PA in every element, obliging 3 dB to 6 dB more PAPR with enhanced PA back off would outcome in a reduction in the efficiency of PA power and decreased output power of array. Nevertheless, the system-level benefits of OFDM modulation in signal processing execution and output in the small range NLOS links [9] can overshadow its transceiver-level shortcomings. The 3GPP Release 15 describes a framework based on OFDM (Table 2).

Table 2 Calculated budget values [1]

	BS-UE downlink	BS-UE uplink
Number of array elements at Tx, Rx	64, 16	16, 64
Total array gain	36 dB	36 dB
Tx EIRP	42 dBm	30 dBm
Rx element sensitivity	−60 dBm	−60 dBm
Rx array sensitivity with margin	−68 dBm	−74 dBm
Maximum acceptable path loss	110 dB	104 dB
Maximum Tx/Rx separation LOS, 3GPP TR38.901, UMi NLOS, 3GPP TR38.901, UMi	87.5 m, 13.7 m	43.9 m, 6.9 m

5 Architecture and Circuit Trade-off

The phased-array element necessitates either a programmable phase shifter or a delay timer to accomplish the beamforming. The phase shifter can be placed either at RF (Radiofrequency) or IF (intermediate frequency) or else in the LO path [12], which overcomes through the whole structure design. It is forthright to observe why RF beamforming is the most repeated alternate for industry solutions, as it eliminates the necessity of separate mixers and corresponding LO distribution while compared with either IF or LO phase shifting mechanism. Additionally, RF beamforming offers three-dimensional blocker elimination upstream of mixer and IF. This lowers the downstream linearity necessities in the system accomplished of detection and placement of spatial nulls at single or multiple blocker positions.

Generally, to carry this process deprived of noteworthy system overhead is a topic of substantial learning because of limited information present at the digital baseband processor about the environment array at any point. So, the system involves some kind of consecutive scanning of the RF beamforming scheme [11]. Besides, the scheme will serve on a very directional scanning strategy of the base station and the blocker positions that influence the whole system potential. Moreover, the null routing through continuous modulus with inadequate resolution phase shifters needs computationally substantial beam-synthesis that creates the rejection of real-time blocker impracticable [12]. The spatial blocker rejection comparable to the sidelobe levels of an array may be anticipated in absence of channel information. It is generally limited to ~10 dB. The ease in mixer and baseband stages in RF beamforming originates in place of lacking phase shifters and summation of millimeter wave. This leads to either a less optimal receiver noise figure or a demanding gain to pay off the phase shifter loss. It is less suitable for MIMO (multiple inputs multiple outputs) applications, in which an array would be capable to direct through a full gain of the array into two or more directions at the same time.

Theoretically, a transceiver should be considered to yield unique manifold beam molded outputs at the RF. This will need extra mixers and hence remove unique profits of RF beamforming. Additionally, splitting of the millimeter-wave signal would be needed to acquire additional gain. An additional method, baseband beamforming, in which all the transceivers and front-end and mixer are fundamentally an individual entity. The signal processing of array happens at baseband in the analog or digital domain. This type of system conserves whole antenna signal data over and done with baseband frequency and offers no opportunity of spatial blocker rejection over receiver. A quantizer is required for each antenna element in the entire digital beamforming receiver. A very high power consumption architecture has been made by the power components of ADC and digital interface. Generally, the transceiver array is extent across physically separated multiple dies from the digital baseband. Therefore it needs more than one extraordinary speed digital buses that connect millimeter wave front end to digital baseband. However, there is the possibility to expand throughput, conforming to power consumption. Uncorrelated ADC quantization noise is inhibited by complete array gain [1]; reduction in the ADC dynamic range by 3 dB for increasing the array elements double to its initial amount.

Digital or analog beamforming at the transceiver expanses to squeezing of the array information, thus improving the output of the interface. Hence, a MIMO system associated with multiple concurrent users in distinctive directions would necessitate a beamformed signal for every user because of this compression. So the throughput of the interface to digital baseband will be therefore scaled by amount of users maintained irrespective of kind of beamforming.

6 Frequency Conversion

Generally, RF beamforming in association with superheterodyne manner is the most common design in the published and industrial applications. In this architecture, the signal is accepted at IF, which is placed in middle of physically separated millimeter-wave transceiver and a baseband integrated circuits. The architecture excludes the vital at IF for the distinct I and Q signaling. This deals with the numerous linked quadrature inequity. Typically in receiver applications [7], the front end response of mmWave offers image rejection at intermediate frequency of the first down conversion stage. This might not be adequate to elude undignified receiver sensitivity, which depends on interference scenario and IF choice.

Additionally, the consideration associated with the super-heterodyne design is required for the resonant gain stages at IF to make the most of the gain. This needs more large inductors than their mmWave complements. Alternatively, direct conversion-based front ends are the foremost break in the recent cellular design. Additionally, not less than a couple of decades, design familiarity is present to defeat their encounters via smart selections of design and calibration. On the other hand, direct conversion design is essentially modest and has no image rejection problem with a set of mixers. Also, its operation necessitates the generation of a quadrature LO at carrier frequency and method of calibration to work with DC offset as well as quadrature imbalance. The additional noteworthy difficulty is that PA may wrench and reduce spectral purity of VCO severely, which is functioning at carrier frequency. The issue can mainly evade by making the operation of VCO at the submultiple of the carrier frequency [13, 17].

Both hybrid and baseband beamforming need LO distribution to several sets of mixers. Distribution of millimeter-wave LO signals to substantially separate dies is a power ravenous attempt. As the routing of the signal is lossy and the buffers require an adequate gain to operate a load of mixer alongside enough swing for good conversion gain. Hence, the designers encounter an architecture alternate: distribution of the LO from a single PLL, or distribution at a lower frequency reference level to the local PLLs located closer to every mixer set to reduce the LO routing loss. This adjustment is dependent on the consumption of power in many local PLLs through the whole die in beamforming process. The power consumption is high in many local PLLs but the uncorrelated noise components of these PLLs are averaged in the calculation of array gain. [19].

7 Front-End Blocks

Transmit power amplifier is liable instead of the concluding amplification and power delivery to antenna with partial alteration, which can harm EVM and spectral emissions. For any system design, the compromises between gain, bandwidth, linearity, and efficiency are present. The advantage of the implementation of phased-array transmitters has advantages of the equal contribution of total output power among the elements in the array, which results in the lowering of output power, which is compulsory for every power amplifier element and hence advancing the linearity. Large power per-element output requiring small arrays and applications may favor SiGe or other similar technologies. Class-AB linear power amplifier is desired topology to enhance the linearity and efficiency in mmWave CMOS Pas [15].

The gain of the device can be improved via the neutralization process, where the intrinsic parasitic gate-to-drain capacitance is nullification approximation has been taken into account. The parasitic feedback path is inhibited via neutralization process; hence it may lead to a more stable device [12]. Compact layout with adequate element to element isolation, suitable gain to reduce receiver noise figure, and low power consumption are the primary requirements for an efficient LNA design. Even though the achievement of minimum probable noise figure is constantly helpful in the system performance, hence, the total prerequisite for the noise figure of a single element can be undisturbed by the factor comparable to array gain [16].

For that purpose, an active mixer can be used, which provides advanced conversion gain with reasonably lower LO swing alongside a good IF-to-RF isolation than a passive mixer. The active mixer has a drawback that it exhausts DC power and undergoes from higher flicker noise, reduced voltage headroom. A passive mixer operating in current mode is commonly used in traditional communication systems because of its better linearity and lower flicker noise, even though it has conversion loss. On the other hand, a passive mixer working in voltage mode can provide extra conversion gain via its sample and hold operation and has the advantage of less flicker noise.

Larger LO voltage swing is required in passive mixers than an active mixer which in turn increases the consumption of power at the LO driver. This limitation can be moderated by using a reduced size mixer switches or by using single-ended topology. Though, these procedures have partial effect at millimeter wave as power loss in local oscillator drive is repeatedly controlled via finite quality factor matching networks. Cautious layout in abridged and symmetric parasitic elements is essential to reduce local oscillator leakage and DC offset for both the mixer topologies, especially when it is embedded in direct conversion transceiver [17].

8 Survey of Integrated Phased Arrays

In recent past, plentiful published works that use a variety of technologies describe the fast growth in inexpensive phased arrays to expedite the fifth-generation market. Kerim et al. in 2018 and Sadhu et al. in 2017 have presented huge 28 GHz phased arrays commissioning of radiofrequency beamforming [10, 11]. Kerim et al. participate four cores into a 0.18 m SiGe front-end chip, which contains the power amplifier, low noise amplifier, antenna switch, phase shifter, and summation blocks. A 32-element array is realized utilizing various front-end chips connected to the PCB. An array is largely owing to the l/2 antenna spacing of 5.4 mm at 28 GHz. This phenomenon keeps front-end blocks near to antennas wherever the routing loss will be having straight influence over the noise figure and power throughput. The LO direction-finding throughout array can be evaded by execution frequency conversion in the distinct IC. The explanation maintenances the measured 1-Gb/s 16-QAM link at 300 meters with 41 dBm output power EIRP at P1 dB [1].

In an alternate approach taken by Sadhu et al., where 32 full transceiver elements are joined on particular 0.13 m SiGe die, with the four-die antenna in a package solution. Highly integrated solution maintenances dual-polarized 64-element arrays to produce 54 dBm P_{sat} EIRP per polarization. Though in Kerim et al. and Sadhu et al., super-heterodyne architectures are employed in SiGe, the effort described by Antonio et al. uses a direct-conversion method in 28 nm CMOS. This approach uses RF beamforming and two-stage injection-locked VCOs. Second stage in the approach is a quadrature injection-locked oscillator (QILO) functioning at 28 GHz [1, 5].

This design conveys an actual reasonable consumption of power and area of each element (50 mW for RX, 85 mW for TX @ 3 dBm, and 0.9 mm^2). In 28 nm CMOS, a completely integrated 4-element transceiver at 60 GHz exploits baseband beamforming alongside the direct conversion. QILO generated quadrature LO signal operating at 24 GHz is multiplied by 2.5 × to the 60 GHz carrier to exclude the voltage controlled oscillator dragging from power amplifier, which can happen with the oscillator operating at fundamental frequency. Single mixer set per antenna element is used in baseband beamforming. The local oscillator signal has been blown out from the single QILO via cascaded local oscillator buffers running at 60 GHz. Alone local oscillator distribution versions 94 mW of power consumption (22% of RX mode power). This explanation comprises the calibration mechanisms for the DC offset impairments and quadrature imbalance as well [1].

9 Scaled CMOS for 73-GHz Design

Multiple challenges are presented during the design of millimeter-wave power amplifier fabricated with scaled CMOS. Interconnection in a scaled version of fabrication enhances back-end capacitance and resistance. This enhancement in resistance and capacitance further lowers the f_{max} & f_t for the devices connected to it. It also reduces the gain achieved and bandwidth. Second, at millimeter-wave frequencies, the device offers lower gain when compared with the RF; if any device at 2 GHz offers maximum gain as 40 dB, then it can provide 8 dB, 14 dB, and 20 dB at 80GHZ, 40 GHz and 20 GHz, respectively. Hence, more than two cascaded stages are required for E band front end having 15–20 dB gain, which results an increase in consumed power [19].

At last, it can be elucidated that both technologies (FinFET as well as FD-SOI) display more thermal dissipation in comparison with bulk CMOS. This results in self-heating where local junction temperature becomes high and it impacts the performance, reliability, and device modeling. Self-heating phenomenon again worsened by the density increment in transistor and finer interconnect metals of the scaled version of nodes. The power consumption should be limited across each device while designing to ensure the reliability of the device. Limited power consumption often leads to point of biasing with f_t and f_{max} values, which are well below the maximum that can be achieved by the process. To demonstrate the scaled version of CMOS nodes suitability in millimeter-wave design, low noise amplifier, and power amplifier was fabricated in 22FFL process of Intel [15]. A stacked amplifier having two-stage and DC-current reuse is facilitated with TMN (transformer matching networks) [15, 17].

The millimeter-wave signal cascaded magnetically with the same transformer when direct current flows via the center tap of the transformer having matched interstage. It lowers the operating voltage and direct current power consumption in every stage is reduced by half for the given current density.

For Fin-FET procedures, the debasement in the performance alongside the abridged drain-source voltage is negligible. The reliability issue because of self-heating is diminished alongside the lessened power consumption. The capacitively neutralized differential pairs are employed as amplifier stages. The bias points of these differential pairs are heightened for a peak obtainable gain and least noise figure though preserving the low power consumption. Transformer-based matching networks in addition to the impedance transformation act as baluns and deliver the ESD safeguard and DC separation. Optimized transformers with maximum available coupling factors are used to enhance the gain and lessen the noise figure. LNA performance measured attains the gain of 20 dB and noise figure fewer than 4 dB at 73.5 GHz at the consumption of power of 10.8 mW.

PA consists of capacitively neutralized transformer similar to the LNA in which the common-source stages are joined through the matching networks. To maximize the backoff efficiency, every stage functions in the class AB mode. To enhance f_t, f_{max}, and current handling of the device, the cautious device layout optimization is used with all the in-place backend connections. Low-k transformers that are generally weakly coupled inductors are employed with interstage matching network to enhance the bandwidth.

Though a general optimization approach is executed over the output matching network to maximize output power delivered and efficiency. The measured peak gain of the PA is 18.6 dB and the bandwidth is 24 GHz. The Psat, OP1dB, and peak PAE measured are +12.8 dBm, +5.7 dBm, and 26.3 percent, respectively.

The calculated PAE at OP1dB is 11.6%, which is maximum in CMOS E-band Pass till the year 2018. Peter et al. simulated this system and calculated an EVM of −26.5 dB for Pout = + 4.7 dBm with 1 dB back off from the P1dB which is measured via AM-AM and AM-PM distortion curves [1]. To aid the results obtained further, modulation experiments were executed alongside the single carrier QAM. The results are produced via an arbitrary waveform generator. To drive the PA, AWG data have been upconverted to a 75 GHz carrier. The output of PA had been downconverted and calculated with a vector signal analyzer (VSA. Even amplitude compression is clear in the constellation plot nonetheless the phase distortion is barely apparent [1].

10 Transceiver at 22 Nm FinFET CMOS Technology

E band having frequency range 71–76 GHz is the required application of the important blocks transceiver. Beamforming is done at baseband with the help of IC using four autonomous direct conversion transceiver cells [1]. Every cell employs its respective quadrature VCO operating at one-third of the carrier frequency and with an integer-N, phase-locked loop locked to a mutual gliding reference. These modulators are made with adjustable gain amplifiers working in current mode. The output of every array element vector modulator is summed up to perform beamforming. The LO signal for every element is produced at the output of the 3 × multiplier at each VCO. Though the present way needs a separate phase-locked loop and a mixer for every element, which possibly increases the area along with the consumption of power. This, in turn, increases the requirement to deal out a millimeter Wave LO signal throughout the die at carrier frequency corresponding to 24 GHz [1, 7].

Additionally, the array that permits the PLLs having high noise and low power averages the phase noise, which is not correlated. Even though the predictable all array gain of the amount 10 log N may not be obtained due to the uncorrelated noise component's secondary effect, which adds the uncertainty in phase relationships among all array components. Phase noise input to EVM & reciprocal mixing is made of the phase noise component, which is not correlated and correlated part that is the same as the central PLL system. This phase noise component is averaged with the help of antenna array. On the other hand, the correlated component is similar to the central PLL system. Offsets in static phase in all LO path are eliminated with the help of phase shifter if necessary.

Thus the power consumption disadvantage in the given architecture may be reduced or eliminated in total by taking benefits of scaled version of phase-locked loop power intake and minimalizing millimeter Wave LO distribution. The reference signal has a comparatively lower frequency (for this case, it is 2.5 GHz). It may be spread to many transceiver units having restricted power intake. In case of LO having its frequency 73 GHz, at 10 MHz frequency, the target phase noise is −110 dBc/Hz. The architecture proposed is quite modular. It can be simply scaled to relatively larger array size. The baseband architecture with beamforming, helps to find the application where the RFPS (phase shifters) would increase the noteworthy loss in signals, such as at advanced carrier frequency or larger fractional bandwidth applications.

Particularly, the additional RF loss at transmitter leads to extra gain stages, which reduce the efficiency and also with its enlarged temperature sensitivity. The beamforming method with baseband may be used in any other band in which the digital beamforming is preferred. Another dissimilar beamforming scheme can be ideal at applications for lower frequency on 39 GHz and 28 GHz, or where the system splitting results in separate front end, as described by Kerim et al., to sidestep the laid-off chip-to-chip interface signals. Band restrictive at the receiver results by the antenna frequency response and front-end matching. Type II 3rd order Chebyshev filter with baseband offers head-to-head channel rejection of 9 dB and it is unified from 1.05–2 GHz. The compensation of the DC offset is essential for any direct conversion ratio at the receiver and transmitter. The compensation is achieved at the baseband by DACs working in current mode and Local succeeding estimation state machine controls the One-bit comparators. Because of having LO available at mixer input and LNA, LO feedthrough is controlled by this compensation at the saturation of receiver baseband. [9, 11].

A main benefit of the suggested architecture is that it can be used in digital beamforming and full chip multiple-input and multiple-output (MIMO) solutions possibly in a scaled version of CMOS technology. An analog baseband PSs (phase shifters) are replaced with data converters and combiners with the digital complex blocks, which are used as mixers [5]. A scaled version of CMOS is quite suitable for digital interfaces having short range, data convertors with high speed and digital processing at low power, the projected architecture is projected to work so that it may be studied whether a digital version of beamforming system can replace its RF beamforming counterpart in terms of power and area. Calibration of phase and gain offsets between the elements is an important drawback in any phased array system. A noteworthy issue in any PAS (phased array system) is a calibration of gain offset and phase among elements. With a central phase-locked loop and mixer relative to an RF beamforming system, baseband beamforming systems are similar to the proposed, which have gain offsets additional to the independent amplifier and mixer blocks [15].

When fine gain offsets are supplied to baseband amplifiers, interelement gain offsets will require factory characterization with the help of test signal and corresponding correction. The noteworthy disadvantage of baseband beamforming tactic is phase offset. The reference phase (path phase) differences among elements that are enlarged by the phase-locked loop frequency multiplication ratio, a factor of 30 for this case. These phase offsets are generally static and these can be remunerated with the calibration with the help of test tones. Real-time beamforming algorithm may treat inter-element PO (phase offsets) in a completely adaptive system made by a digital baseband like in the digital beamforming. It may not require an obvious routine of calibration. [12].

Similar to every quadrature system, calibration is required for PO (phase offsets) and I/Q gain. The PS (phase shifters) with baseband having file bit resolution, the bandwidth of 2 GHz along with 40 dB dynamic range consumes 6 Milliwatt for each element. PS (phase shifter) power with baseband must be in the order of same for compensation in gain from a PPS (Passive PS) when comparing with Radio Frequency beamforming. The gain stage using compensation approximately 10 Milliwatt, RFPS (phase shifter) may lose more than 10 dB at 73 GHz [1]. The set for such array has four transceivers RFICs (working in E band) for a 16-element array, with sizeable packets so that these may be located one after another to produce a random array size. I/Q signals with baseband are clubbed and divided on package with the help of RSN (resistive split networks).

11 Conclusion

In past decades, the short-range applications with tremendously huge-bandwidth communications have ascended and these are well supported by millimeter-wave band spectrum. There are some architectural and circuit challenges, which are focused all over the physics related to amplification and distribution of signals 10 times in the frequency range in comparison to the 4G cellular signals. Generally, a large number of antenna arrays are required to deliver a practical aperture for antenna, with the small physical size of the antenna array at higher frequency.

Author Contributions
This book chapter is written through the contribution of all authors. All authors have been approved the final version of the chapter. All authors declare no competing financial interest.

Acknowledgements Author Shalu C. is grateful to NCflexE, IIT Kanpur for providing encouragement and financial assistance to carry out the research work. Amitesh Kumar is thankful to Department of Electrical Engineering, National Institute of Technology, Patna for providing resources to write this book chapter. Amitesh Kumar is thankful to the Council of Scientific & Industrial Research to provide research fellowship to carry out research during his stay at the Indian Institute of Technology, Indore.

References

1. Sagazio P, Callender S, Shin W, Orhan O, Pellerano S, Hull C (2018) Architecture and circuit choices for 5G millimeter-wave beamforming transceivers. IEEE Commun Mag 56(12):186–192
2. Josefsson L, Persson P (2006) Conformal array antenna theory and design. Wiley, USA, pp 1–9
3. Rappaport TS, Xing Y, MacCartney GR, Molisch AF, Mellios E, Zhang J (2017) Overview of millimeter wave communications for fifth-generation (5G) wireless networks—with a focus on propagation models. IEEE Trans Antennas Propag 65(12):6213–6230
4. Ghosh A, Thomas TA, Cudak MC, Ratasuk R, Moorut P, Vook FW, Rappaport TS, MacCartney GR, Sun S, Nie S (2014) Millimeter-wave enhanced local area systems: a high-data-rate approach for future wireless networks. IEEE J Sel Areas Commun 32(6):1152–1163

5. Buzzi S, D'Andrea C, Foggi T, Ugolini A, Colavolpe G (2018) Single-carrier modulation versus OFDM for millimeter-wave wireless MIMO. IEEE Trans Commun 66(3):1335–1348
6. Kutty S, Sen D (2016) Beamforming for millimeter wave communications: an inclusive survey. IEEE Commun Surv Tutor 18(2):949–973
7. Barati CN, Hosseini SA, Mezzavilla M, Korakis T, Panwar SS, Rangan S, Zorzi M (2016) Initial access in millimeter wave cellular systems. IEEE Trans Wireless Commun 15(12):7926–7940
8. Kummer WH (1992) Basic array theory. Proc IEEE 80(1):127–140
9. Mo J, Alkhateeb A, Abu-Surra S, Heath RW (2017) Hybrid architectures with few-bit ADC receivers: achievable rates and energy-rate tradeoffs. IEEE Trans Wireless Commun 16(4):2274–2287
10. Kibaroglu K, Sayginer M, Rebeiz GM (2018) A low-cost scalable 32-element 28-GHz phased array transceiver for 5G communication links based on a 2 × 2 beamformer flip-chip unit cell. IEEE J Solid-State Circuits 53(5):1260–1274
11. Sadhu B, Tousi Y, Hallin J, Sahl S, Reynolds SK, Renström O, Sjögren K, Haapalahti O, Mazor N, Bokinge B, Weibull G, Bengtsson H, Carlinger A, Westesson E, Thillberg JE, Rexberg L, Yeck M, Gu X, Ferriss M, Liu D, Friedman D, Garcia AV (2017) A 28-GHz 32-element TRX phased-array IC with concurrent dual-polarized operation and orthogonal phase and gain Control for 5G communications. IEEE J Solid-State Circuits 52(12):3373–3391
12. Puglielli A, Townley A, LaCaille G, Milovanović V, Lu P, Trotskovsky K, Whitcombe A, Narevsky N, Wright G, Courtade T, Alon E, Nikolić B, Niknejad AM (2016) Design of energy- and cost-efficient massive MIMO arrays. Proc IEEE 104(3):586–606
13. Bai T, Alkhateeb A, Heath RW (2014) Coverage and capacity of millimeter-wave cellular networks. IEEE Commun Mag 52(9):70–77
14. Heath RW, González-Prelcic N, Rangan S, Roh W, Sayeed AM (2016) An overview of signal processing techniques for millimeter wave MIMO systems. IEEE J Sel Topics Signal Process 10(3):436–453
15. Yong SK, Chong C (2006) An overview of multigigabit wireless through millimeter wave technology: Potentials and technical challenges. EURASIP J Wireless Commun Netw 2007:1–10
16. Yang H, Herben MHAJ, Akkermans IJAG, Smulders PFM (2008) Impact analysis of directional antennas and multiantenna beamformers on radio transmission. IEEE Trans Veh Technol 57(3):1695–1707
17. Marzetta TL (2010) Noncooperative cellular wireless with unlimited numbers of base station antennas. IEEE Trans Wirel Commun 9(11):3590–3600
18. Hoydis J, ten Brink S, Debbah M (2013) Massive MIMO in the UL/DL of cellular networks: How many antennas do we need? IEEE J Sel Areas Commun 31(2):160–171
19. Huang X, Guo YJ, Zhang A, Dyadyuk V (2012) A multi-gigabit microwave backhaul. IEEE Commun Mag 50(3):122–129

Low Noise Amplifiers Designing

S. Sakthi Arun, Sangeeta Singh, and Rajeev Kumar Arya

Abstract In this chapter, the basic designing and characteristics of the LNA for 5G network are discussed in depth. LNA designs are specially considered for various 5 G applications. Its designing approach varies with the narrowband and wideband over the band of frequency interested. Here various topologies to achieve the better optimized performance in the circuits are covered. Generally, the LNAs at front end of receiver which receiver the input signal from the antenna and amplifiers it with less noise. In LNA, the gain and the noise are the most importance consideration factors for the design. This chapter covers different noise figures for single-phase and multistage amplifiers are demonstrated. Further various challenges faced during the 5G circuits designing are also covered. Along with the many popular LNA circuits for 5G, cascode limitations and scaling effects, neutralized CS amplifier and LNA design at 22 nm and 14 nm FinFET architectures are also discussed along with the in-depth designing features.

Keywords 5G · LNA · FinFET · 22 nm FinFET designing

1 Low-Noise Amplifier

In recent year CMOS technology is widely implemented due to its advantages of cost reduction less supply voltage and highly integrated systems on the chip. Scaling in CMOS technology provides reduction in gate delay doubles the device density and reduces the energy per transitions. Generally for RF front end circuits design of low voltage CMOS analog circuits is highly demanded. In Radio frequency integrated circuits (RFICs) amplifiers are most important components. In RF receivers the Low Noise Amplifiers is the first and the most critical block. LNA is the first amplification stage of any receiver block hence it plays a vital role in the receiver side.

S. Sakthi Arun · S. Singh (✉) · R. K. Arya
Department of Electronics and Communication Engineering, National Institute of Technology Patna, Patna, Bihar, India
e-mail: sangeeta.singh@nitp.ac.in

S. Singh et al. (eds.), *CMOS Analog IC Design for 5G and Beyond*, Lecture Notes in Electrical Engineering 719, https://doi.org/10.1007/978-981-15-9865-4_4

2 Introduction

In the 5G world, the data rates are generally greater than 1 GB/s. Due to the wide applications of Mobile internet, IOT, Augmented reality, virtual reality, remote computing, e-health services, automotive driving, and indoor hotspot, dense urban the numbers of user cases are increased dramatically. Basically, Nanoscale CMOS transistors performance is degraded due to the electrical stress, so when these devices are used in radio frequency application, it reduces the circuit performance. LNA designs are specially considered for various applications. Its' design varies with the narrowband and wideband over the band of frequency interested. There are various topologies were followed to achieve the better optimized performance in the circuits. In general, the LNAs at front end of receiver which receiver the input signal from the antenna and amplifiers it with less noise. In LNA, the gain and the noise are the most importance consideration factors for the design. Apart from these factors, some more are considered such as linearity, which represents the stability of the system without the distortion and interference because due to the large number of input signal, the system has the chance to be nonlinearity. As this is the first block, the input matching is the primary factor to be implemented as a first cut design. The proper input matching provides the maximum power transfer.

The overall Noise figure (NF) of the LNA is defined as the signal to Noise ratio at the output side to the signal to Noise ratio at the input. As per the IEEE standards IEEE 802.11b, the frequency band of 2.4 GHz is widely used for the general application such as hotspot, WLAN or any public overcrowded officers where the number of users is very high. Hence in 5G, the frequency spectrum is provided for wideband. Generally, the wideband low noise amplifiers are one of the key components in wideband receiving systems. The performance of the wideband LNA is greatly significant by the receiver noise figure and the high sensitivity receiver. Considering the application of W-band imaging system, which is greatly temperature sensitivity and resolution over the wideband interested, this directly relates the Noise Figure. For the multimode communication, the LNA has some specifications such as better bandwidth, high gain, linearity, proper input matching and low noise figure. If multiple LNA designs is considered it not only increases the area but also degrades the gain, bandwidth and noise performance due to the additional parasitic also the power efficient is reduced. So considering the entire above factor, the LNA design methodologies are briefed in this chapter.

3 Noise Figure Basics

Noise is one of the undesirable factors, which are added in the RF system. In RF Circuits, noise along with the nonlinear distorted signal bounded to affect the system performance. In RF front end process Amplifiers, filters and mixers are subjected to the input noise as same as like the input signal. The various components in amplifiers are the source for the excess noise-generating factor. Practically, all the amplifiers are subject to noise.

The noise in the system is measured as Noise figure or Noise factor. Basically, they are defined as the ratio of signal to noise power at the input to the signal to noise power at the output. The perfect noise-free amplifier maintains the signal to noise ratio at its input as well as output.

$$F = \frac{SNR_i}{SNR_o}$$

where SNR_i is the signal to noise at the input and SNR_o is the signal to noise at the output.

But the noise figure increases if there is a decrease or degradation of signal to noise ratio. The low noise figure represents the low noise is added by the network. The basic fundamental explanation of noise is given by H. Nyquist, W. Schottky and J.B. Johnson. In this topic, we will discuss further about the various source of noise.

3.1 Thermal Noise

The thermal noise is generated due to the random thermal motion of the electrons. As the source of this noise is due to the thermal motion of electrons, it depends upon the absolute temperature T, hence they are directly proportional to it. When temperature T becomes zero, the thermal noise value becomes Zero [1].

The representation of thermal noise due to resistor R is given by the series voltage generator $\overline{v^2}$ as shown in Fig. 1a. Similarly, the shunt current generator $\overline{i^2}$ due to R is shown in Fig. 1b.

$$\overline{v^2} = 4KTR\Delta f$$

$$\overline{i^2} = 4KT\frac{1}{R}\Delta f$$

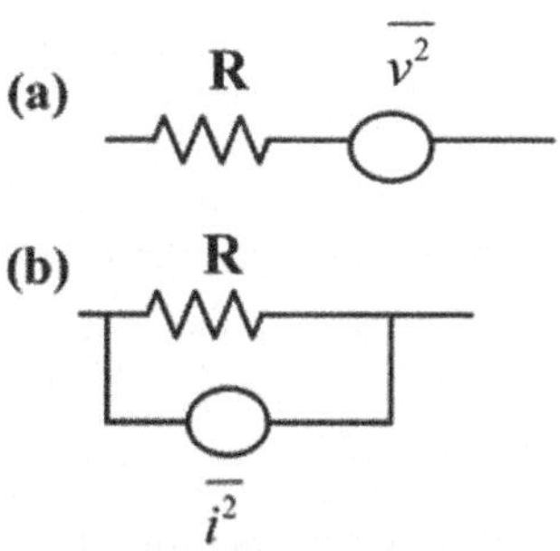

Fig. 1 Representation of Thermal Noise [1]

where k is Boltzmann's constant where $4KT = 1.66 \times 10^{-20}$ V-C. We can say the noise spectral density is independent of frequency and it's applicable till 10^{13} Hz. R^2

White Noise that is the another noise occurs due to the thermal noise from 1 and 2, we can derive the Norton equivalent.

$$\overline{i^2} = \frac{\overline{v^2}}{\mathrm{R}^2}$$

Thermal Noise is present in every linear passive resistor, which includes the conventional resistance and the radiation resistance of antennas, loudspeakers and microphones.

Generally in loudspeakers and microphones, the noise occurs due to the thermal motion of the air molecules whereas in antennas is due to the black body radiation of object at which the antenna is directed.

The above equations represent the thermal noise value of monolithic and thin film resistors, whereas carbon resistors used as external component in IC also provide flicker Noise.

3.1.1 Noise Calculation of Resistor's in Series

Considering the Resistors, R1 and R2 are connected in series and their noise generated noise is given by

$$\overline{v1^2} = 4KTR_1\Delta f$$
$$\overline{v2^2} = 4KTR_2\Delta f$$

The overall noise is calculated by adding the two individual noise generated by R_1 and R_2.

$$v\mathrm{T(t)} = v_1(\mathrm{t}) + v_2(\mathrm{t})$$

Hence

$$\overline{v\mathrm{T(t)}^2} = \overline{[v_1(t) + v_2(t)]^2}$$
$$= \overline{v_1(t)^2} + \overline{v_2(t)^2} + \overline{2v_1(t)v_2(t)}$$

As mean square product value is Zero, the equation becomes

$$\overline{v\mathrm{T(t)}^2} = \overline{v_1(t)^2} + \overline{v_2(t)^2}$$

Substituting value from 6 and 7 in 11, hence $\overline{v_\mathrm{T}(\mathrm{t})^2}$ becomes

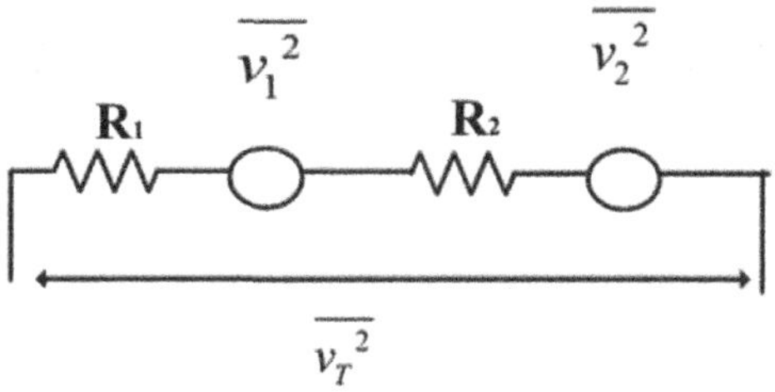

Fig. 2 Circuit for the calculation of the total noise $\overline{\upsilon T2}$ produced by two resistors in series. [1]

$$\overline{\upsilon \mathrm{T(t)}^2} = 4KT(R_1 + R_2)\Delta f$$

In general, thermal noise is often known as Johnson noise or Nyquist noise. The plot of $S(f)$ versus frequency is shown in Fig. 2a. The Sinusoidal current RMS value i is shown in Fig. 2b. As the signal is sinusoidal, its calculation is conventional method. In case of multiple sources, then each source is individually calculated and added the individual mean contribution. This is applicable to all individual sources except gate induced noise.

Considering the circuit shown in below Fig. 2 where R1 and R2 are connected in series and their noise generated is given by

$$\overline{\upsilon 1^2} = 4KTR_1\Delta f$$
$$\overline{\upsilon 2^2} = 4KTR_2\Delta f$$

3.2 *Flicker Noise*

Flicker Noise occurs due to the charge traps in Si/Sio_2 interface because of contamination and crystal defects and they are released in random fashion and time constant, which results in the source for noise signal. In general, these noises occur in all active devices and also in carbon resistors [2].

In active devices, the flicker noise is associated with the direct current, whereas in carbon resistors there is no flicker noise as long as no direct current is passed through it and it acts like normal noiseless resistors. This type of resistors is mostly used in low frequency integrated circuits. In case of high frequency, a metal film resistor that contains no flicker noise is used.

Carbon composition resistors exhibit current dependent noise because of the random formation and also due to the loss of macro arc among neighboring carbon granules. MOSFET generally generates more Flicker noise as compared with the other active device because of the smoothing of fluctuations in channel charge due to the larger gate oxide capacitance.

The model for flicker noise is given by

$$\overline{in^2} = \frac{K}{f}\frac{gm^2}{WLCox^2}\Delta f$$

where

K is device-specific constant

g_m is transconductance

W is width of device

L is Length of device

Cox is gate oxide capacitance per unit area.

In general, smaller K is obtained from P-MOSFETs as their channel is buried. The flicker noise is also known 1/f as its spectral density is inversely proportional to the frequency.

3.2.1 Techniques to Reduce Flicker Noise

There are some basic techniques to reduce the flicker noise.

(i) As the flicker noise is also inversely proportional to gate area, increasing the size of gate area at the expense of gate capacitance require higher Transconductance to bring down the corner frequency to acceptable value.
(ii) Correlation of double sampling and chopper stabilization eliminate flicker (1/f) noise.

3.2.2 Shot Noise

Shot noise was first discovered by Schottky in 1918. Shot noise generally occurs due to direct current flow in active devices like diodes, MOS transistors and bipolar transistors.

The Shot noise occurs when the hole from P-region and electron from the n-region crosses the junction and when crosses they diffuse away the entire minorities.

The fluctuation in I and diode current I_D is termed a shot noise and in general known as mean square variation and given by

$$\overline{\overline{i^2} = (\mathrm{I} - \mathrm{ID})^2}$$

$$= \lim_{T\to\infty}\frac{1}{T}\int_0^T(\mathrm{I} - \mathrm{ID})^2 dt$$

The resulting noise current is given by

$$\overline{i^2} == 2qI_D\Delta f$$

where q is electronic charge (1.6×10^{-19} C) and Δf is bandwidth in hertz.

This mean square value gives the noise current value. But the equation is valid until it is comparable to 1/T where T is carrier transit time through depletion region.

Two condition required for Shot Noise:

(i) Flow of direct current.
(ii) Potential barrier over which the carrier extracted.

Shot Noise is a mixture of thermal and generation and recombination. The generation and recombination noise caused due to the emission and capture of charge carriers by traps and they mainly depend on temperature, frequency and biasing.

3.2.3 Noise in Nonideal Capacitor and Inductor

In integrated circuits, the capacitors are used for some specific function or they are formed a parasitic. Similarly, inductors are used in high frequency communication. The ideal inductor and capacitors generate no noise whereas in case of nonideal capacitors and inductors in integrated circuits contains the series resistance, which generate noise.

4 MOSFET Noise Parameters

The channel that forms due to the gate source voltage is resistive and exhibits the thermal noise when the drain current flows and it contributes major source for noise in MOSFET. The source of noise is Flicker noise due to traps of charge carriers near Si/SiO_2.

The drain current noise is a combination of thermal noise and Flicker Noise.

$$\overline{id^2} = 4KT(\frac{2}{3}g_m)\Delta f + K\frac{ID^a}{f}\Delta f$$

where 4KT ($\frac{2}{3}$ g_m) Δf is Thermal Noise, K $\frac{IDa}{f}$ Δf is Flicker Noise, I_D is Drian bias current, K is constant in device, a is constant value range from 0.5 and 2, g_m is transconductance.

The above equation is valid for long channel, whereas in case of short channel, the above noise value will be 2–5 times greater, which leads to the increase of hot electrons thermal noise.

Due to the gate leakage current, the MOS transistor generates shot noise, which is the another source of noise and it is represented by $\overline{ig^2}$.

$$\overline{ig^2} = 2qI_G\Delta f$$

The above $\overline{ig^2}$ is negligible small due to the low gate current.

At radio frequency due to the random components between the gate and channel, results in the random fluctuation along the channel, which leads to the thermal noise, are given by below equation:

$$\overline{ig^2} = \frac{16}{15}KTR\omega^2C_{gs}^2\Delta f$$

where $C_{gs} = \left(\frac{2}{3}C_{OX}WL\right)$.

For short channel, both the gate current is added each other and correlated with the thermal noise, which gives the total noise current in MOS transistor.

4.1 Thermal Noise in MOSFET's

The model for MOSFETs drain current noise, induced gate current noise and their cross-correlation coefficient are given by

$$\overline{id^2} = 4KT\Delta f\gamma g_{d0}$$

where γ value is 2/3 when channel pinched off and 1 when drain biased zero.

$\gamma \rightarrow$ bias dependent factor

K $\rightarrow$ drain output conductance under zero drain bias. The gate induced current noise is

$$\overline{ig^2} = 4KT\Delta f\delta g_g$$

where,

$\delta \rightarrow$ bias dependent factor

$g_g \rightarrow$ real part gate to source admittance and it is given by

$$g_g = \frac{\omega 2Cgs2}{gd0}$$

4.2 Design Noise Figure

In RF front end amplifier, filter and mixers contribute their own noise along with the input signal noise. In case V_s is the input signal applied to the amplifier, the SNR is represented for relating the signal with noise to understand the system noise-related metrics. If S_i is the input signal and N_i is the noise along with input signal and if the amplifier is noise free then the input signal is amplified by the power gain of the amplifier (G). Then the output signal $S_o = GS_i$ and the output noise $N_o = GN_i$, hence the overall output $SNR_o = S_o/N_o = SNR_i$. In the nonideal case of amplifier, the amplifier that made up of component generates noise, which is indicated as N_e. Hence the total outputs noise is given by $N_o = GN_i + N_e$. N_e is measured in terms of signal to Noise ratio (SNR) called Noise Factor (F). Figure 3 shows the noise and two ports: (a) amplifier; (b) amplifier with excess noise; and (c) noisy two-port network.

The Noise Factor (F) is given by

$$F = \frac{SNR_i}{SNR_o}$$

In ideal case, $SNR_o = SNR_i$ then

$$F = 1$$

In nonideal case, i.e., if the amplifier is noisy then SNR_o is less than SNR_i and F > 1.

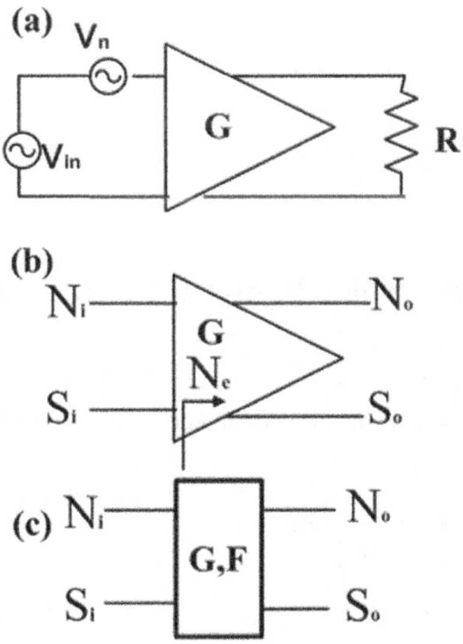

Fig. 3 Noise and two ports: **a** amplifier; **b** amplifier with excess noise; and **c** noisy two-port network. [3]

$$F = \frac{S_i}{N_i}\frac{N_o}{S_o}$$

In case of noisy N_o is given by $N_o = GN_i + N_e$, hence the Noise Factor is

$$F = \frac{S_i}{N_i} + \frac{GN_i + N_e}{GS_i}$$

$$F = 1 + \frac{N_e}{GN_i}$$

In standard circuits, the input signal contains the internal resistance, which contributes the input noise along with the input signal over the Bandwidth (B) used and it is given by

$$N_i = N_R = KTB$$

where K is Boltzmann constant and T is absolute temperature (290 K).

If the input is connected to their resistor then the Noise Factor becomes

$$F = 1 + \frac{Ne}{GKT_0B}$$

Noise Factor is the most important parameter to determine the noise performance of a system.

Noise Figure is measured as Noise Factor in decibels and represented as

$$NF = 10\log_{10}F$$
$$NF = SNR_i(dB) - SNR_o(dB)$$

In case of cascaded system, the Noise factor is given by calculating the noise of the first stage and then followed by the cascaded system.

5 Noise Figure for Multistage Amplifiers

Figure 4 demonstrates the Cascaded noisy two ports. In RF cascaded system, the output noise power is expressed with and without noise contributed by amplifier as

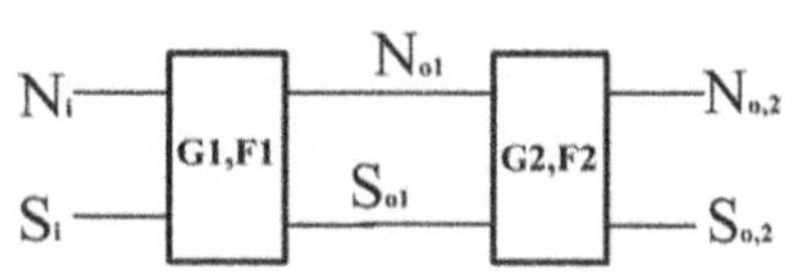

Fig. 4 Cascaded noisy two-ports. [3]

$$N_o = GKT_oB$$

With N_e,

$$\begin{aligned} N_o &= GKT_oB + N_e \\ &= GKT_oB\left(1 + \frac{Ne}{GKT0B}\right) \\ &= FGKT_oB \\ N_e &= (F - 1)GKT_oB \end{aligned}$$

Hence

Let us calculate the noise in the second stage of the amplifier.

$$N_{e,2} = (F_2 - 1)KT_oBG_2$$

The total Noise power at the output of second stage is given by

$$\begin{aligned} N_{o,2} &= (F_2 - 1)KT_oBG_2 + N_{o,1}G_2 \\ &= (F_2 - 1)KT_oBG_2 + F_1KT_oBG_1G_2 \end{aligned}$$

For m stage, the output noise is given by

$$No,\, m = \sum_{n=2}^{m}[(Fn-1)KToB\prod_{i=2}^{n}Gi] + F1KToB\prod_{n=1}^{m}Gn$$

The total noise factor

$$F^T = N_{o,m}/\left(G^TN_{i,1}\right)$$

Where G^T is total cascaded available gain and $N_{i,1}$ noise power input to first stage.

The overall Noise Factor is expressed as

$$F^T = F_1 + \frac{F_2 - 1}{G_1} + \frac{F_3 - 1}{G_1G_2} + \frac{F_4 - 1}{G_1G_2G_3}$$

That is represented as $FT = F1 + \sum_{n=2}^{m} \frac{Fn-1}{\prod_{i=n}^{m} Gi-1}$

The above equation is known as Frii's formula.

6 Challenges 5G Circuits

6.1 Linearity

Wideband CMOS LNAs cover the wideband of standard frequency bands. CMOS had certain advantages such as low cost and less area on chip RF systems and good performance over the frequency of operation. WB LNAs should contain some good performance such as proper input and output matching, Low Noise Figure, high sensitivity, high gain, good linearity and low power consumption. Less area, cost and voltage were the other concerns. Linearity is the big challenge to the designer due to the interferers and intermodulation/cross modulations due to blockers. Some of the Linearization techniques enhance the linearity as follows.

A. Predistortion technique

The Schematic of differential capacitor cross couple Common gate (CCC-CG) circuit using predistortion technique [4] enhances the linearity by using the auxiliary transistor to operate in the weak or moderate inversion region, which cancels out the nonlinearity of the transconductance stage. This technique not only provides better linearity IIP3 but also proper input matching circuit with gain enhanced and noise reduction.

B. Postdistortion Technique

B.1 Active Feedback

Postdistortion is the another technique, which cancels out the distortion at the output nodes by setting the low input impedance at DC and $2f_{RF}$. The circuit diagram of postlinearization technique proposed for two-stage amplifier structure.

Circuit Design

The circuit implants two stages [4]. The first stage includes CG and CD as cascode structure. The common gate due to its low input impedance helps for the wideband matching and off-chip balun, which tunes the output capacitance.

Common Drain (CD) cancels the noise and distortion at the output nodes OUTP and OUTN. The second stage comprises of Common Source (CS) and CD amplifiers, which utilizes Cascode structure. This enhances the gain of the first stage and cancels out the noise of first stage and distortion at the output nodes. The circuit drives the large capacitive loads at the output with minimum gain attenuation at higher frequencies due to its low impedance. The noise generated by CG passes through the differential path and gets cancel due to the same polarity. When the frequency increases the IIP3 of the circuit reduces. Basically in wideband applications, linearity mainly depends on the frequency between two tones rather than the narrowband.

C. Cross-connected Differential pairs

In broadband application, the high IIP2 helps the LNA to stay linear in the presence of interference far from the fundamental tone. The cross-connected differential pair technique improves both the input second- and third-order intercepts points (IIP2 and IIP3) by canceling all the intermodulation components from the output current. The common mode current generated from the LNA with opposite polarity by cascading the two differential pairs with cross-connected output and these are eliminated at the output due to the opposite polarity and it further enhances the linearity. This technique improves the noise performance by canceling the thermal noise of the input and the auxiliary transistor at the output.

Circuit Description

The design of UWB CMOS LNA [4] in which the transistor utilizes Common Gate, which provides better wideband matching. In addition, the Cascode structure provides better isolation and gain. The additional transistor cancels the common mode Intermodulation components. The noise cancellation path improves the noise performance. The additional transistor attenuates the IM2 and IM3 current of the transistors and eliminates the third-order nonlinear current of the circuit. It results in enhancing the linearity and reducing the NF. The major challenging in this process is operation of the auxiliary transistor in different operating regions.

D. Multiple Gate Transistors

Derivative superposition or multiple gate transistors-based techniques [4] improve the IIP3 but the factor IIP2 is degraded. The gate induced noise degrades the noise figure. Hence the LNA factors such as High gain, low noise figure and good matching are difficult to attain by this method. To overcome these issues, Complementary DS is introduced, which contains NMOS/PMOS pair to improve the IIP3 without degrading the IIP2. Complementary DS improves the linearity. Differential topology improves the IIP2 and the stacked NMOS/PMOS circuit enhances the IIP3. By using the current mirror in the circuit, the on-chip inductors are eliminated to achieve the compact design.

7 Popular LNA Circuits for 5G

7.1 Linearity Enhancement Techniques

Due to the increase in the demands for the broadband high-speed wireless communications system in millimeter-wave (MMW) frequency bands for high speed data requires linearity over a wideband of frequency. But due to parasitic components of high frequency degrades, the nonlinearity parameter for enhancing the linearity [5]. In LNA, linearity and the Noise Figure is the main factors. The linearity has increased by enlarging the size of the device with additional DC power consumption, which further enlarges the parasitic capacitance C_{gs}, which degrades the Noise Figure.

$$NF_{min} = 1 + 2\pi f K Cgs\sqrt{\frac{(\mathrm{Rg} + \mathrm{Rs})}{\mathrm{Gm}}}$$

where *Cgs*, R_g, R_s, G_m *and K* are the gate source capacitance, gate resistor, source resistor, transconductance of the transistors and constant value, respectively. Therefore, improving LNA linearity without increasing the size of the device is another major challenge.

There are some linearity techniques such as: (i) Optimal bias technique in which biasing the transistor for the IM3 free region reduces the gain and also biasing the gate for linearization is sensitive to temperature and process variation. (ii) Active postdistortion (PD) method biases all the transistor to strong inversion to achieve insensitive matching and less temperature variation. (iii) Derivative superposition (DS) technique nulls the nonlinearity for LNA. The major drawback of all the above three techniques is effective till 10 GHz and fails in millimeter wave frequency without affecting the Noise Figure.

To overcome these challenges, the parasitic-insensitive linearization technique has been implemented, which includes the body biased postdistortion method, which utilizes the gate bulk bias tuning, which adjust the amplitude and phase of nonlinearity at higher frequency parasitic also source-sensed DS technique with distributed elements to eliminate the MMW parasitic capacitors.

7.2 *Parasitic-Insensitive Linearization LNA*

Linearization improves the linearity, data rate and spectral efficient in a receiver. As the frequency of the operation increases, the parasitic degrades the nonlinearity rejection. Parasitic-Insensitive Linearization is insensitive to the process variation and high frequency parasitic variation. This method includes (1) Distributed CS stage and (2) body-biased PD.

A. Distributed DS Technique for Common-Source LNA

To overcome the issue of nonlinearity rejection without affecting the other nonlinearity parameter such as gain, NF and adding extra power consumption. In high frequency operation as the DS linearization performance is affected due to parasitic capacitors. The source-sensed DS technique together with distributed network eliminates the impact of parasitic due to MMW capacitors.

Common Source LNA with distributed CS technique [5]. The distributed circuit elements around the parallel transistor tune out the parasitic capacitors. At the weak inversion region, device has positive g_m and at strong inversion it has negative character. The size of transistors should be properly chosen. Reducing the size of transistor, increases the length of the source degeneration line and vice versa. Hence it results in process variation. Lines around the linearizer create additional phase delay of nonlinear terms, which need to be optimized for better IMD_3 performance. Microstrip lines have replaced passive components to achieve a compact chip area and minimize the effect of process variation.

B. Body-biased PD Technique for cascode LNA

The schematic of two-stage Cascode LNA with body-biased PD technique [5]. The additional inductor added to cancel out the parasitic capacitances.

The third-order term g_{m3} is the main contribution for the IM_3 in RF systems. The selection of transistor size can result in the minimization of third-order nonlinear terms. Biasing in the weak inversion region for IM_3 optimization and at strong inversion region for parasitic insensitivity and optimizing IM_3. Simultaneously, transistor is forward body biased, which further reduces V_{th} and results in the parasitic insensitive linearization.

7.3 *Noise Canceling Technique*

Noise Canceling Technique is widely used for wideband matching network in LNA circuit. The noise generated from input matching transistor is reduced by feed-forward path while input impedance also matched simultaneously. This technique provides good wideband input matching, low Noise Figure (NF), sufficient voltage gains and high linearity [6]. To attain the high voltage gain, the voltage amplifier is added while maintaining the Noise canceling and input matching simultaneously. Therefore, Figure of Merit (FOM) of the Noise canceling LNA is improved.

The functional block of LNA consists of:

1. An Amplifier stage providing the source impedance matching, $Z_{in} = R_s$.
2. An Auxiliary amplifier sensing the voltages (Signal and Noise) across the input source.
3. A combining network that combines the output of two network, providing the result of canceling the noise generated by input stage and sum up the amplified voltage signal.

8 Cascode Limitations and Scaling Effects

8.1 *Cascode Limitations*

Cascode topology is widely accepted in Low Noise Amplifier design due to its advantages like impedance matching, gain and stability. But the Cascode topology design attains certain limitations such as follows:

A. Effect of Width Variation in Noise

Increasing the width of the Cascode devices increases the shielding from output and also reduces the value of L_S for certain Z_{in}.

Increasing the width of the Cascode device certainly affects the overall noise performance of the device. As the width of the Cascode stage increases, the generated noise power from the Cascode stage also increases and simultaneously increases the noise resistance. Hence the different width of the Cascode stage different noise match condition as well as different value of noise resistance. In case width increases, then the Cascode stage further increases the 40% extra noise power in input stage, which in turn increases the NF by 0.5 dB.

B. Limitation in Voltage Headroom

The addition of cascode device limits the voltage headroom due to stacking of two transistors and also rises in noise contribution. The Cascode topology consumes additional voltage headroom but limits the transistor biasing hence raising their noise contribution.

C. Complexity Design

In Cascode topology usually forward body biasing technique handles the issue of voltage head room. The conventional Cascode required low power consumption, but as Common Gate is stacked with top of Common Source which required high voltage supply for transistor operations. But the issue of this technique is increased in the complexity of its design as it requires extra supply voltages [7].

The conventional forward bosy bias technique needs an additional bias circuit to supply the bulk terminal of the MOSFET for obtaining the forward bulk source bias.

D. Increase in Chip size

The Cascode structure consists of CS and CG popular for mmWave LNA due to its high gain and compact size. As the frequency goes higher CG device in the cascode device in Cascode will causes the extra noises at its output, hence it results in trade-off between gain and noise [8].

At higher frequencies, parasitic capacitance causes the drain noise of the cascode device to contribute noise, which further increases the Noise Figure. To cancel out the parasitic capacitance noise effect, a parallel resonant inductor L_P is placed to resonant with C_X. But the introduction of inductors occupies larger areas in chip.

E. Linearity Limitations

Cascode reduces the drain to source voltage, which degrades the linearity because of low supply voltage supply operation.

8.2 Scaling Effects

In recent technology, Scaling paves the way for high speed and low power design. However, small in transistor size faces problems such as reliability degradations, process variability, aging effects and field variations.

Due to this process variation, circuit faces deviation from measured results. The problems due to process variability are Local or intra-die and global or inter-die effects [9].

The aging effect results in reliability degradation due to hot carrier injection (HCI), mobility degradation, gate dielectric breakdown and negative bias temperature instability (NBTI). Similarly, field variations also contribute Short Channel Effects (SCE) such as Drain Induced Barrier Lowering (DIBL), velocity saturation and mobility degradation, which occurs when the electric field crosses the critical field value.

DIBL occurs due to the high drain voltage, which lowers the barrier height between the source and the channel region results in reduction of threshold voltage. DIBL is modeled as:

$$V_t' = V_t - V_{ds}$$

where η is DIBL coefficient and V_t' is changed threshold voltage due to change in V_{ds}. DIBL effect results in increase of subthreshold current due to which transistor gets turned ON prematurely. The field variations degrade the mobility, which results in change in β value which is given by

$$\beta = \mu C_{ox}\left(\frac{W}{L}\right)$$

finally all these changes due to process variation, aging and field variations occurring in RF IC circuits results in change in threshold voltage, transconductance and noise parameters of transistors. The threshold voltage is given by

$$V_t = V_{t0} + \gamma\left(\sqrt{(|2\phi F| + VSB)} - \sqrt{(|2\phi F|}\right)$$

where V_t is threshold voltage with substrate bias, V_{SB} is source-body bias voltage, V_{t0} is threshold voltage when V_{SB} is zero, $2\phi_F$ is surface potential and γ is body effect parameter is given by

$$\gamma = \frac{tox}{\varepsilon ox}\left(\sqrt{2q\varepsilon siNA}\right)$$

where t_{ox} is oxide thickness, ε_{ox} is oxide permittivity, ε_{si} is permittivity of silicon, N_A is a doping concentration and q is charge. The variation causes changes γ and temperature (T), which further causes fluctuations in $2\phi_F$ causes shift in V_t. Due to this g_m gets affected due to this process variation also results in noise signals. In application of Network on chip (NOC), it is designed with re-routing option for providing the reliability of the device by providing the runtime monitoring and counter measures for compensation of reliability errors, but it consumes more area and power.

In recent scaling technologies, Scaling the gate oxide scaling is the challenge to analog circuits as the gate oxide reaches the single atom size [10]. To attain this, it needed a higher K gate dielectric which increases the flicker noise and parasitic series resistances.

9 Neutralized CS Amplifier

In the recent days, the wideband application such as automotive radar, millimeter-wave imaging and high data rate communication is growing rapidly. To improve gain with better matching and low noise figure over the band interested is the basic requirement for LNA [11]. This Neutralized technique improves maximum available gain and stability over the wideband performance.

9.1 Circuit Implementation

The figure contains the three stage of implementation in which first two stages being single-ended cascode and the third stage is differential cascode. To reduce the power consumption and interference from the antenna, the single-ended cascode in the first and second stages. Between the second and third stage a low-k balun is inserted to convert the single-ended signal to differential for neutralizing the signal. The third stage includes neutralized CS and a neutralized common-gate (CG) with peaking inductors.

Finally, the LNA drives the double-balanced mixer, this technique is widely used in RF/mm-wave receivers. The effect of gate resistance on f_{max} and NF can be reduced by optimizing the width of the transistor fingers and connecting the multiple fingers in ring structures. The better transistor width and current density provide optimized results of noise matching and impedance matching. The inductors and transformers are designed to show the optimized value of quality factor by using the metals.

9.2 Design Procedure

The input impedance of the network is given by

$$Z_{IN} = (1 - k)L_1 + KL_1||[(1 - K)L_2 + Z_X]$$

where k is the coupling factor, L_1 and L_2 are the self-inductance of primary and secondary windings and Z_X is the input impedance of M_2

The simulated S_{11} shows that the transformer-based matching provides the better matching than the conventional LC network as it has the multiple poles and achieves wideband matching over 70–90 GHz. The impedance is given by

$$ZX \approx \frac{1}{CGD1s} + LS1s + \frac{g_{m1}L_{s1}}{CGS1}\left(1 - \frac{2CGD1}{CGS1}\right)$$

where gm1, C_{GS1} and C_{GD1} are the transconductance, the gate to source capacitance and the gate to drain capacitance. The inductor values are chosen such that it resonates with the parasitic capacitances and improves the bandwidth matching over the wideband of frequencies. The center frequencies over the three stages with the low-k balun and double neutralized LNA improve the gain.

10 Double-Neutralized Technique

The third stage LNA utilizes a differential cascode doubles neutralized CS LNA, which boost the gain and stability [11]. The parasitic capacitance can also be cancelled by insertion of Metal-Oxide-Metal (MOM) capacitors between the drain and source of differential pair. Neutralizing both the CS and CG enhances the stability and gain. In addition, CG also extends the output swing, which enhances the linearity for supply voltage 1.8 V. It is necessary to prevent the circuits from power stress during the start of supply 1.8 V.

The Neutralizing CG in the double Neutralized circuit provides extra Poles and Zeros, which extends the bandwidth. Comparing with the conventional neutralized and single neutralized CS amplifier, double neutralized CS amplifier enhances the stability and maximum gain by properly selecting the value of capacitors. The improper selection of capacitors values reduces the sensitivity and stability due to the process variation.

11 FINFET—A Design Perspective View

The continuation scaling in CMOS (Complementary Metal Oxide Semiconductor) technology results in Short channel effect, which creates several problems, which are discussed in earlier topics. Hence to overcome these MOSFET problems FinFET-based tri-gate, double gate, four gate, all around gate technology are introduced for better performance, the schematic of FinFET structure.

The gate has the better control over the channel, which occurs between Source and drain. The I_{ON} and I_{OFF} are improved by this device structure.

Some of the main advantages of FinFET:

1. The Short Channel effect due to Scaling can be suppressed by thin Si fin.
2. Reduction of parasitic resistance and improved current drive.
3. The better gate control provides the better performance over V_{th}.
4. The larger the fins larger the device size and current flow.
5. Lower off state current, lower supply voltage required for device operation, hence it results in low power consumption.

11.1 Challenges in MmWave Design

In mmWave design, the FinFET-based technology faces some of the challenges in the design as [12].

A. Limited available Gain:

The presence of gate resistance (R_g) and capacitance(C_{gs}) forms low pass network, which limits the gain of the device at higher frequencies. At mmWave frequencies, usage of multistage amplifiers to attain the necessary gain may result in increasing power consumption and eventually lower the bandwidth.

B. Self-Heating Effect (SHE):

The increase in the transistor density and current density can increase the temperature around a device, which impacts reliability of device and local interconnects. In additional, the temperature dependent parameter such as threshold voltage and mobility degrades, hence the power consumed across the device should be limited. FinFET maintains strong drive strength and channel control at low supply voltages.

C. Parasitic of Interconnects:

Scaling of interconnects leads to increase in parasitic, which degrades the performance of RF circuits by limiting its f_t and f_{max}.

11.2 LNA Design at 22 nm FinFET Architecture

To overcome the problem of high power consumption and high area, the LNA design at 22 nm FinFET architecture is preferred. This topology effectively reduces the DC power consumption by half for each amplifier stage due to its current reuse technology hence resulting in a power efficient LNA design.

The FinFET-based LNA topologies have reduced Drain Induced Barrier Lowering (DIBL) and aggravated Self-heating effect (SHE). This reduces the supply voltage to lesser heat generation in device and better amplifier performance circuit.

11.3 Design Procedure

The schematic of the stacked two-stage differential LNA with DC current reuse [13].

This circuit contains two stages as neutralized differential pairs as amplifier and transformer as matching network, which separates the DC bias connections and enables the current reuse, which effectively reduces the power consumption.

11.4 Effect of Pad Capacitances

The pad capacitances act as an input matching element and the inductor components provide better output matching network. The neutralization technique that discussed in earlier topic proves that it increases the stability and maximum gain of the network. To achieve the ideal performance of neutralized differential pairs (NDPs), the neutralized capacitor is designed using metal interconnect parasitic capacitance. Self-Neutralized amplifier with different width is constructed by arranging the multiple NDP units, which reduces layout and modeling overhead by using this modular and connect-by-abutment.

11.5 Impact of Transformer

Using transformer, benefits for impedance matching networks with low coupling factor for wideband. Although a transformer benefits for low K, it may suffer insertion loss in practical IC design due to its quality factor. The maximum efficiency of transformer can be expressed as:

$$\eta\text{max} = \frac{1}{1 + 2\sqrt{(1 + \frac{1}{(\text{QmK})2})\frac{1}{(\text{QmK})2}} + \frac{2}{(\text{QmK})2}}$$

where $\text{Qm} = \sqrt{\text{QpQs}}$ is the geometric mean of quality factors of primary and secondary coils, K is the coupling factor of two coils.

Basically, in LNA design, transformer required higher K values to achieve high gain and lower noise figure over the desired bandwidth. In this topology of neutralized differential pair and transformer matching networks, a compact and power efficient design is achieved with lowest noise figure and lowest power consumption.

11.6 FinFET-Based Intel's 22FFL Process

In communication system over a higher frequencies of operation, FinFET-based Intel's 22FFL design offers a high density, low leakage and energy efficient mmWave system [12].

11.7 Noise Performance

In FinFET, noise performance is mainly analyzed in a receiver and oscillator design. Intel's 22FFL FinFET NF_{min} is below 2 dB up to 73 GHz, this can enhance the sensitivity of the receiver and also it achieves the superior flicker noise performance over planar.

11.7.1 FinFET-Based LNA Implementation

RF-CMOS applications provide high speed and low power devices with reduced area consumption with increasing mask cost of scaled silicon. To overcome the scaling effect in deep submicron technologies, SOI FinFET-based LNA design approach is implemented [14].

A. 40 nm FinFET Inductor and Inductor less LNA Architecture

Scaling of CMOS beyond 45 nm needs an additional design to have a better control over the reduced channel. FinFET-based LNA design shows a better RF performance. The inductor based and inductor less based FinFET-based LNA for narrowband and wideband applications which allow for better matching network and noise-free design [14].

Scaling down the transistor size, the gate loses the control over the channel due to Short Channel Effects. Hence to overcome these challenges, the FinFET is introduced, which has the gate all around the channel. The minimum physical gate length of both LNA is 40 nm.

While FinFET attains higher gain over the short channel and the similar gain is achieved by planar device for similar f_T [14]. Moderate gain and f_T lie in both the devices. Typically lowering the noise requires increasing current. For inductively degeneration LNA, the noise figure (NF) is proportional to f_0/f_T, where f_0 is the operating frequency. Similarly, feedback LNA possesses the NF proportional to g_m^{-1}. It is clear that FinFET suffers from lower f_T and lower bandwidth.

The FinFET device attains much higher NF_{min} due to source and drain series resistance (R_S and R_D) in FinFET due to the extension of source and drain also the parasitic capacitance. The gate resistance is also higher in FinFET due to larger fin spacing in device structure. For better performance, the optimal layout design is required.

B. 16 nm FinFET Architecture

In wireless sensor network for IOT application ultra-low power (ULP) radio frequency, integrated circuits are key challenges to battery life as they are efficient to operate for the low voltage supply. To attain the ULP & ULV, LNA is designed to operate at subthreshold region on gate and triode region on drain terminal and also stable region of VSWR should be larger than 10:1.

The modified version of LNA [15] which can operate in 100 mV ULV. To have the better stability, the design implants as below:

(i) Source and load impedances need to be properly selected.
(ii) Voltage gain is 0.8 dB less.
(iii) V_{DD} greatly reduced by 3x.

11.8 Region of Operation

To achieve a high transistor efficiency of g_m/P_{DC}, the subthreshold region is adopted in gate and triode region in drain side. By applying the gate side subthreshold region, the efficiency enhanced by 12% also the drain side triode approach boosts the g_m/P_{DC} to 100% under 100 mV operation. The better stability is achieved. The transistor speed is enhanced by enlarging the gate width which increases the f_T with 5× of operating frequency.

11.9 Measured Result

The modified LNA achieves Power Gain of 6.9 dB and NF 3 dB at 2.4 GHz with V_{DD} of 100 mV.

C. Inductor-less 16 nm FinFET Low Power LNA Architecture

In earlier FinFET-based LNA includes the inductor in the design for the high performance of LNA but the usage of multiple on-chip inductor occupies the huge silicon area with increased cost design, which leads to high interference signal and power consumption. Hence design of small area, i.e., inductor less design provides a compact design for SOC design [16]. But it results in challenge of noise matching at input and gain degradation due to capacitance loading at output. To have the better noise figure and gain, the PMOS is stacked with the superior g_m and design utilizes current reuse technique.

11.10 Linearity Optimization

Due to low V_{DD} structure, linearity or IIP3 becomes more using complementary superposition higher linearity is achieved at low supply voltage of 0.4 V. By adjusting the gate bias of PMOS and NMOS, their complementary gm_3 curve is superposed for zero gm_3.

11.11 Measured Results

For the supply voltage of 0.4 V, the LNA achieves Noise Figure as 2.5 dB, −6 dBm in band IIP3 and 35 dB gain over 20 MHz Channel Bandwidth. The area occupied is very less as compared with earlier inductor based design.

12 LNA for 5G Circuit Using 14 nm FinFET CMOS

FinFET-based LNA provides better gain and low power consumption with less area occupied in chip. In this section, we discuss some of the LNA design using 14 nm FinFET-based design as follows.

A. Ka-band Phase shifting LNA

Millimeter-wave phase control components are essential to applications such as electronic beam forming, transceiver phase array system for adjusting the phase of each antenna and beams. RF beam forming consists of RF phase shifters with low power solution. The passive phase shifter provides better linearity with low DC power consumption with high insertion loss, which need to be compensated in design and also face area issues due to on chip inductors [17]. Hence it is not suitable for the area efficiency phase shifter in receiver.

To overcome such issues due to passive shifters, active shifters using transistor-based design is preferred, which provides high integration level with better gain and accuracy under constraint power budget. However, LNA design with better Noise Figure is required. In this section, LNA design using 14 nm FinFET-based CMOS phase shifter is discussed, which need to produce minimize loss, chip size and gain variation while achieving at 22.5 degree of phase shift.

12.1 Circuit Design

The schematic of Ka-band phase shifting Low Noise Amplifier [17] circuit composed of two stages (1) Conventional cascode FET for more gain and Noise Fig. 2 Triple-stacked FET for phase shifting operation. The gate finger and fin number of core 14 nm FinFET CMOS are 120 and 6 respectively with gate width 54 μm and current consumption is 6.2 mA at V_{GS} of 0.5 V. The V_{DD1} of first stage is 1 V and V_{DD2} of second stage is 1.5 V is maintained to overcome the voltage headroom of triple-stacked FET. The phase shifting operation is done by control of C_k. The layout size of the ka-band phase shifting LNA is 900 μm × 550 μm.

The triple well isolation used for CG-FET. Pwell and Nwell is connected to source and V_{DD}. The gate of the stacked CG-FET is terminated with variable capacitance array.

The voltage gain is shown below by ignoring the drain-to-source capacitance (C_{gd})

$$A_V = \frac{V_o}{V_x} = \frac{g'_{ds} + K_{gm}}{g'_{ds} + g_L}$$

where $K = \frac{C_K}{C_{gs} + C_K}$.

The variation of C_k in above equation can also affect the gain. To minimize this variation g_L should be large for better phase shifting. The triple stacked FET with gate terminated capacitance provides lower load impedance to 2nd stage. Similarly, higher g_L minimize the gain variation. Moreover, additional stacked CG-FET operated as isolation buffer to provide higher gain due to higher output impedance.

12.2 Measured Results

The gain is around 15 dB and variation is less than 2 dB due to low input impedance of stacked CG-FET and phase control renage is 22.5°. The fixed input and output characteristic guarantee stable phase shifting operation in phase array receiver system. Phase shifting LNA provides better phase error compensating solution in phased array receiver system.

Author Contributions
This book chapter is written through contributions of all authors. All authors have given approval to the final version of the manuscript. The authors declare no competing financial interest.

Acknowledgements Authors are thankful to Department of Electronics and Communication Engineering, National Institute of Technology, Patna for providing resources to write this book chapter.

References

1. Gray PR (2009) Analysis and design of analog integrated circuits, 5th edn
2. Goo J-S (2001) High-frequency noise in CMOS low-noise amplifiers Stanford University, p 149
3. Steer M (2010) Microwave and RF design: a system approach, vol 3
4. Mohebi Z, Parandin F, Shama F, Hazeri A (2020) Highly linear wide band low noise amplifiers: a literature review (2010–2018). Microelectronics J 95:104673
5. Li WT, Tsai JH, Yang HY, Chou WH, Gea SB, Lu HC, Huang TW (2012) Parasitic-insensitive linearization methods for 60-GHz 90-nm CMOS LNAs. IEEE Trans Microw Theory Technol 60:2512–2523
6. Chen KH, Liu SI (2012) Inductorless wideband CMOS low-noise amplifiers using noise-canceling technique. IEEE Trans Circuits Syst I Regul Pap 59:305–314
7. Li C, Li M, He K, Tarng J (2010) A low-power self-forward-body-bias CMOS LNA for 3–6.5-GHz UWB Receivers IEEE Microw Wirel Components Lett 20(2):100–102
8. Huang BJ, Lin KY, Wang H (2009) Millimeter-wave low power and miniature CMOS multicascode low-noise amplifiers with noise reduction topology. IEEE Trans Microw Theory Technol 57:3049–3059
9. Balamurugan K, Kartha KS (2018) Design of LNA using body biasing techniques for enhanced reliabilty and robustness. In: Proceedings of the 3rd international conference on electronics information systems ICCES 2018, pp 803–810
10. Ponton D, Palestri P, Esseni D, Selmi L, Tiebout M, Parvais B, Šiprak D, Knoblinger G (2009) Design of ultra-wideband low-noise amplifiers in 45-nm CMOS technology: comparison between planar bulk and SOI FinFET devices. IEEE Trans Circuits Syst I Regul Pap 56:920–932
11. Pan D, Duan Z, Chakraborty S, Sun L, Gui P (2019) A 60-90-GHz CMOS double-neutralized LNA technology with 6.3-dB NF and -10dBm P-1dB. IEEE Microw Wirel Components Lett 29:489–491
12. Callender S, Shin W, Lee HJ, Pellerano S, Hull C (2018) FinFET for mm Wave—Technology and circuit design challenges. In: 2018 IEEE BiCMOS compound semicondutors integrated circuits technology symposium BCICTS 2018, pp 168–73
13. Shin W, Callender S, Pellerano S, Hull C (2018) A compact 75 GHz LNA with 20 dB Gain and 4 dB noise figure in 22 nm FinFET CMOS technology. In: Digest of technical papers. IEEE radio frequency integrated circuits symposium, pp 284–287
14. Borremans J, Parvais B, Dehan M, Thijs S, Wambacq P, Mercha A, Kuijk M, Carchon G, Decoutere S (2008) Perspective of RF design in future planar and FinFET CMOS In: Digest of technical papers. IEEE radio frequency integrated circuits symposium, pp 75–78
15. Lu Y T and Jin J De 2016 100-mV 44-μW 2.4-GHz LNA in 16 nm FinFET technology *IEEE MTT-S Int. Microw. Symp. Dig.* **2016-August** 1–3
16. Yeh EHV, Lo AH, Chen WS, Yeh TJ, Chen M (2017) A 16 nm FinFET 0.4 V inductor-less cellular receiver front-end with 10 mW Ultra-low power and 0.31 mm2 ultra-small area for 5G system in Sub-6 GHz band
17. Kim Y, Jang P, Lim J, Ko W, Heo S, Lee J, Cho T B (2018) A Ka-band phase shifting low noise amplifier with gain error compensation for 5G RF beam forming array using 14 nm FinFET CMOS. In: Proceedings of the IEEE international symposium on circuits systems, May, pp 1–4

Receiver Architectures for 5G: Current Status and Future Prospects

Amitesh Kumar, Brajendra Singh Sengar, Shalu Chaudhary, Saurabh Kumar Pandey, Sushil Kumar Pandey, Md. Hasan Raza Ansari, and Aaryashree

Abstract In this chapter, the recent progress in receiver architecture and various aspects of the available receiver architectures have been discussed. Besides, an overview of the systematic classification of architecture has been analyzed. Documentation of new possibilities and system-level trade has been closely inspected. Certainly, there is a requirement of low-power, flexible, and high-performance receiver architecture for the successful implementation of the 5G network. Different works in this regard have been considered as examples for discussing the status and prospects of architectures with respect to 5G future. Various architectures considered in this chapter can be very valuable to design 5G network in future and will expose the research community with new possibilties to explore for further improvements.

A. Kumar (✉)
Nextgen Adaptive Systems Laboratory (NASL), Department of Electrical Engineering, National Institute of Technology Patna, Patna, India
e-mail: amitesh.ee@nitp.ac.in

B. S. Sengar (✉)
Department of Energy Science and Technology, Centre for Advanced Studies, Lucknow, India
e-mail: brajendra@cas.res.in

S. Chaudhary
National Centre for Flexible Electronics, Indian Institute of Technology Kanpur, Kanpur, India

S. K. Pandey
Department of Electrical Engineering, Indian Institute of Technology Patna, Patna, India

S. K. Pandey
Department of Electronics and Communication Engineering, National Institute of Technology Karnataka Surathkal, Mangalore, India

Md. Hasan Raza Ansari
Department of Electronic Engineering, College of IT Convergence Engineering, Gachon University, Seongnam-si 13120, South Korea

Aaryashree
Department of Applied Chemistry, Shibaura Institute of Technology, 3 Chome-7-5 Toyosu, Koto City, Tokyo 135-8548, Japan

S. Singh et al. (eds.), *CMOS Analog IC Design for 5G and Beyond*, Lecture Notes in Electrical Engineering 719, https://doi.org/10.1007/978-981-15-9865-4_5

Keywords Wireless communication · 5G · Receiver architectures · RF mixer architecture · Baseband TIA · I/Q generation

1 Introduction

The receiver architectures need a lot more consideration as we are moving to the era of 5G. Among the important architectures existing today, we need to consider their applicability and prospects. The correct choice of receiver architecture for a specified system is very crucial as it will accompany the requirements of Low Noise Amplifiers (LNAs), duplexers, Mixers, filters, and Analog-to-Digital (A/D) Converters (ADCs), which is very vital to realize a broad, flexible and low power system. Besides, 5G is also about receiving a high-frequency signal and reconstructing data, hence it would require some changes in the physical layer. The antenna would always be there at input and data at the output, which requires a receiver for down-conversion and analog-to-digital conversion. There are several works in this regard [1–8]. Various scholars have time to time put forward different developments in architecture for the innovations. Here we consider some of these works to decide the interrelation between different architectural features and corresponding performance parameters.

2 Mixer Architectures for 5G Network

Mixers, which are part of transmitter, are deployed to decode the signal band from one frequency to another. While transmitting through the transmitter, the low frequency is reproduced with the oscillator signal to produce the RF signal and vice versa on the receiver side, *i.e.*, RF signal is reproduced with the carrier signal, to restore the original low-frequency signal) [9]. Wireless receiver architecture is a vital part of a mobile receiver design process. For any given system specification, the choice of duplexers, filters, LNAs, Mixers, and ADCs depends upon the type of wireless receiver architecture. The preferred architecture should be specific to fulfill the requirement of power consumption, cost, and integration capabilities. Hence making room for any new approach such as 5G would require many changes in the given system. Considering the most common receiver architectures, authors [10] have illustrated a proposed framework based on the designs of common receiver architectures as shown. It demonstrates the degrees of freedom while outlining a receiver architecture. Next to elucidate the assignment of the three-dimensional framework architectures, two-dimensional predictions are there [10].

Table 1 Following are the characteristics of IEEE 802.15.4 down-converter [11]

Noise figure (NF)	<20.5 dB
IIP3	> −32 dBm (high gain) > −10 dBm (low gain)
Input Power	− 85 to −20 dBm
SFDR	>38.5 dB
IIP2	>10.5 dBm
Phase Noise	−73.5 dBc/Hz (5 MHz offset)

3 The RF Mixer Architecture and the Power Splitter

The present technology of scaled down communication devices puts rigorous aims of excellent operation parameters in the small silicon space. The stipulations for the wireless scheme operation parameters according to IEEE 802.15.4 are listed in Table 1 [11]. Gladson et al. [1], have considered planning and realizing, small noise, largely linear and excellent gain RF down-converter networks for IEEE 802.15.4 (LR-WPAN) implementation.

The work goals to realize low noise figure, high gain, and high gain concurrently through allocating the circuit performance among various phases of the RF down-converter (RFDC) [1]. The input step of the orthodox mixer detects the voltage, whereas the three-stage frequency alteration circuit detects the feedback current. Though, in the orthodox mixing phase, input match is difficult to realize while, in this RFDC, resistance may be assigned as $1/g_m$ of the transistor considered. Besides, in orthodox mixers, the considered load is inert whereas in this RFDC [1], the load is substituted by additional gain phase. Therefore, the considered RFDC [1] could deliver improved gain as well as decreased noise figure (NF). In the considered three-stage circuit by Gladson et al., the RF signal is detected at the source for the NMOS CG devices T1 & T2. V_b is gate bias for the transistors (common gate). The transistors MT1 and MT2 offer the required bias for transistors T1 and T2. V_{cs} delivers the required bias for the transistors MT1 and MT2. The RF input is administered by the TIA phase, whereas the signals from the TIA transistors are served to the mixing.

Table 2 [1] displays the assessment of the considered current sensing based three-stage RFDC compared to the down-converters stated in the literature [12–19].

4 IF Mixer

Vigilante et al. have reported the IF-based architecture in the following work. IF mixer, IF transconductor, together with baseband amplifier (transimpedance). As can be observed, a current approach is selected to exploit the linearity of previous stage (tailback of RX chain). Further, a *n:* 1 transformer has been used to interface

Table 2 Various CG-based mixers compared as follows [1]; a—high gain, b—low gain; PLS—Post Layout simulation

Reference	[12]	[13]	[14]	[15]	[16]	[17]	[18]	[19]	[1]
Technology (nm)	130	180	65	180	130	130	180	180	180
DSB-NF (dB)	5.95	8.8	2.16–3.9	10	<15	7.2	4.2–6.2	8.5	12.76
PLS/measured	PLS	Meas.	Sim.	PLS	Meas.	PLS	Meas.	Meas.	PLS
Area (mm^2)	1.397	1.04	0.21	2.08	0.75	1.76	0.11	1.41	0.0043
IIP3 (dBm)	−11	−5.8	5.1	−18.8	−44a/−13.5b	−10.8	−12a	−33.5a	−3.43
Conversion gain (dB)	22.5	19	25.6	18.2	35.4a/34.7b	22.3	28.1a	45.9a	23.69
Power (mW)	9	24.5	10.8	5.57	33	0.905	5.2	1.42	12.6 (including VCO)

the transconductor, as well as the switching quad in the RF track. The TIA has been applied with a two-stage OTA (differential one) using resistive feedback. It offers ≈50-dB gain, with ≈50 input impedance near ≈5.6-GHz BW (−3 dB), whereas using ≈8.46 mA supplied by 0.9 V source. Further, the TIA is monitored through on-chip buffers to initiate the 50 measurement tools [20].

To generate I/Q (LO IF) signals, two dividers (inductor-less) built on CML static latches sealed by a source has been used. Static dividers have been favored in contrast to injection-locked ones owing to superior locking range as well as lesser silicon space, at the cost of greater power consumption [20].

5 Baseband TIA and I/Q Generation

Further to consider other receiver designs, we have the work of Bhat et al. [21], which displays numerous receiver architectures (mixer-first) in the perspective of in-band linearity. The receiver considered in this work does not accomplish high in-band linearity owing to noteworthy signal fluctuation at TIA input. It can reach good in-band linearity, however it is noisier owing to the committed 50Ω similar resistor. To contemplate the noise-canceling topology, where linearity is restricted by the LNTA for working at RF frequencies. There is quite alternative way to intensify in-band linearity, for its inefficiency of input matching which is not realistic in numerous operations [21].

The noise-canceling architecture is proposed by authors [21] such that every active circuit is at baseband frequencies and input matching is supplied by *RB*, as impedance is tuned by frequency to the input through passive mixer. A supplementary route through LNTA with the transconductance G_m annuls the noise for resistor. The filter (N-route) considered through the source impedance and capacitor castoffs OoB interferers. The receiver architecture realizes better in-band linearity due to the ground (virtual) at TIAs' input and the LNTA functioning in baseband.

There is LNTA and its' biasing circuit displayed in the work [21] with the inverter-only three-stage OpAmp considered for 6 GHz UGB. It is comprised of three cascaded inverters in the leading route and a inverter with feed-forward route. The foremost route provides enough gain as it reaches 175 MHz so as to accomplish the prerequisite linearity. A grouping of the Miller compensation along with removal of right half (plane zero) and a feed-forward high-frequency inverter route alleviates the OpAmp. Also, the circuit variables are there for the OpAmp-A2. To consider OpAmp-A1, the values have been altered to scale to attain the essential loop gain.

Further to talk about an architecture of I/Q upconverter, a corresponding design is by Nam et al. [22]. The considered Tx upconversion sequence is comprised of an I/Q double-balanced mixer (passive one), IF amplifier, a current-operated GM stage, a resistor bank with digital capacitor array for altering the mid frequency, and a single-ended output balun (transformer-based). A negative *gm* (NGM) cell applied has been linked in parallel with the GM phase output, which drives a feeble reversal to abandon the second derivative of *gm* that primarily adds to IM3. Furthermore, a super-source follower with confined feedback with M3 has been deployed to take forward the low IF input resistance for the mixer (passive). Also, it delivers a high impedance load to adaptable gain amplifier working in voltage mode. Besides, the passive mixer (double-balanced) as depicted is engaged to achieve an upconversion with increased linearity and small flicker noise [22].

6 Measurement Results

Here we discuss the measurement results of the work from respective architectures. First, we discuss the RF down-converter as proposed by the author [1] from the corresponding work [1]. To discuss the consistency of the RF down-converter, variation analysis related to corner, voltage and temperature are performed for the prerequisite range of frequency for conversion gain, noise figure [1]. Monte-Carlo simulation for 500 samples with the maximum deviation (3σ) is used for device mismatch evaluation for the down-conversion network considered for analysis of various parameters such as gain, IIP2, NF, IIP3, and phase noise of the VCO for 5 MHz offset and 100 MHz. The Monte-Carlo simulation observations (the standard deviation (3σ) value and mean (μ) value are analyzed [1].

To discuss the results of the architecture of the work, as discussed by authors [20], a LNA test chip is there to discuss the chip magnetic coupling (parasitic) between transformers in multistage amplifiers. The conversion gain and the input match compared with frequency are considered as well as the NF. Measurements and simulations show similarity, particularly for the high frequency with the extensive bandpass bandwidth. The mentioned IIP3, as evaluated by using two interferers at 84-GHz and 1-GHz offset is considered. Table 3 displays the outcomes and equates the results to 70/80-GHz CMOS LNAs [20].

Table 3 Different LNAs IN 70/80-GHz BANDS reported in Literaure [20]

Ref.	[20]		[23]		[7]	[24]
CMOS Tech. (nm)	28		28		65	65
VDD (V)	0.9		0.9		1.2	1
Gain (dB)	29.6	18	23.8	19.3	17.5	9.4
P_{DC} (mW)	31.3	11.7	30.6		30	9.7
BW_{-3dB} (GHz)	28.3	30.7	10		2	15
ICP_{1dB} (dBm)	−28.1	−12.3	−18.5	−15	−22	n.a.
GBW(THz)	0.85	0.24	0.15	0.09	0.01	0.04
NF (dB)	6.4 − 8.2	7.8 − 9.8	4.9	5.6	7.4	6.7
f_c (GHz)	82.3	81.1	79		77	79
FOM (dB)	18.2	12.3	7.7	3.2	−19	1.9

Table 4 Literature comparision between various works [21]

Reference	[21]	[25]	[26]	[27]	[28]	[29]
Architecture	BBNC	Mixer-first	FTNC	Mixer-first	ZIF RX	ZIF RX
f_{RF} (GHz)	1–6	0.4–3.5	0.08–2.7	0.2–8	0.4–4	0.4–6
Power (mW)	172	38–75	35.1–78	$50 + 30 \times f_{LO}$	200	6600
Area (mm^2)	0.48	0.23	1.2	0.8	49	68.7
IF bandwidth (MHz)	175	50	2	10	50	100
In-band IIP3 (dBm)	9	7	– 3	– 10	– 4	12
NF(dB)	2.5–5	2.4–2.6	1.9	2.3–5.4	2	12
Technology	22 FDX	28 nm	40 nm	45 nm SOI	45 nm	65 nm

Further, the test setup to measure the receiver architectures of the work by Bhat et al. [21] has been discussed. The measurements have been accomplished through a source (single-ended) with a passive balun for the receiver. IIP3 is evaluated through 2-tone tests for both in-band as well as out-of-band (OoB). For in-band evaluations, $f1$ and $f2$ are at $\Delta f - 2\ MHz$ and $\Delta f + 2\ MHz$ correspondingly. To consider the OoB IIP3, two tones $f1$ and $f2$ are at Δf and $2\Delta f - 50\ MHz$ so as IM3 products display the evaluated IM3 curve (for $\Delta f = 20$ MHz). The considered measurement reveals that the IIP3 is effective till -20dBm of input power. As we may see, which displays the flexibility of receiver for multiband as well as multichannel process. It displays the evaluated NF, SC21, and IIP3 at 50 MHz offset and S11 through Local Oscillator (LO) frequencies (1–6 GHz). NF and IIP3 maintain 3 dB and 10dBm correspondingly for LO sweep. Table 4 displays TIA bandwidth and other performance factors of the general receivers and compare it with the performance of receiver. The considered receiver has high IIP3 with 12dBm band-edge IIP3.

Then, we discuss the results of Nam et al. [22] with the I/Q upconverter chip block (0.58×0.98 mm^2). The output spectrum measured and 2-tone IF inputs (100 MHz as well as 110 MHz) of (a) one with and (b) other without stimulating the NGM cell has been considered. The OIP3 evaluated with NGM is around 14.45 dBm, while OIP3 not considering NGM is merely 11.87 dBm to display +2.58 dB enhancement due to the third harmonic elimination consuming insignificant extra power of 438 μW. Whereas, as the NGM is there, the considered insertion loss is less (< 1 dB). Further, for the evaluation of the IRR (Image Rejection Ratio) frequency (300 MHz) is applied with IF input at the IF input port. The LO leakage measured and IRR are −37.7 and −46.9 dBc, correspondingly. There is a mismatch of gain measured for I/Q channels by changing the LO frequency for static IF frequency (PIF = − 30dBm at fIF = 300 MHz) < 0.2 dB. Considering gain mismatch and IRR, it may be estimated that the mixer phase mismatch is less than 0.5° [22]. Table 5 draws the comparison between the instigated mixer with other upconversion mixers [30–32]. Here, the considered mixer deploying QSG in the upconverter displayed the minimum chip size.

Table 5 Literature comparison of various upconversion mixers [22]

Reference	[22]	[30]	[31]	[32]
Technology	65 nm	65 nm	65 nm	0.13um
Chip Area (mm^2)	0.57**	0.42**	1.23**	0.86**
RF Freq. (GHz)	10–11.5	60	27.6–28.4	23.4–29.2
OIP3 (dBm)	14.45	N/A	15.7	N/A
Gain	9.12	6.12	11.4	−1.9
OP_{1dB} (dBm)	2.65	−7.3	2	0.3
LO Freq. (GHz)	20–23	51–69	28	18.9–29
P_{LO} (dBm)	~10	0	−13	0
P_{DC} (mW)	48.7**	17.8*	15*	39.3**

* Mixer Only, ** Mixer with LO buffer

7 Conclusion

In conclusion, we have witnessed many receiver architectures is given, each with their pros and cons along with their variations. With 5G approaching in upcoming technological era, a low-power, flexible, and high-performance solution is needed. Various architectures have been considered in this chapter. A novel three-stage RF down-converter as described above can be there for highly efficient, low-power, low-cost, area-efficient requirements, which can be applied for wearable, portable 5G applications (2.4 GHz band [1]. Further, other discussed design for LNA [20] shows improvement over conventional designs and can be used as the first broadband RX for *E*-band deep-scaled bulk CMOS point-to-point communication links [20]. Besides, another architecture, as the considered receiver [21] can accomplish higher linearity (in-band) for a large range of RF frequency range (1–6 GHz). Owing to noise-rejecting characteristics, input match is significant with truncated NF [21]. Thereafter, an I/Q upconversion series has been considered in 65-nm CMOS procedure [22]. The designed upconverter takes up an very small area of 0.57 mm^2 smallest in comparison to the other conventional upconversion mixers [22]. These above architectures can be very useful to work for a 5G network in future and will open new dimensions for further applications. The above works can be very useful to foresee the world with 5G. Their applications and implementation would definitely be very crucial to understand and tackle several challenges in implementing the 5G.

Author Contributions

This book chapter is written through contributions of all authors. Amitesh Kumar and Brajendra Singh Sengar have worked together on this publication and contributed equally. All authors have given approval to the final version of the manuscript. The authors declare no competing financial interest.

Acknowledgements Amitesh Kumar is thankful to Department of Electrical Engineering, National Institute of Technology, Patna for providing resources to write this book chapter. Amitesh Kumar is thankful to Council of Scientific and Industrial Research to provide research fellowship to carry out research during his stay at Indian Institute of Technology, Indore.

References

1. Bhaskar SCGM (2020)A 3-stage RF down-converter network for portable, wearable 5G applications. SN Appl Sci
2. Bhagavatula V, Zhang T, Suvarna AR, Rudell JC (2016) An ultra-wideband if millimeter-wave receiver with a 20 GHz channel bandwidth using gain-equalized transformers. IEEE J Solid-State Circuits 51(2):323–331
3. Bassi M, Zhao J, Bevilacqua A, Ghilioni A, Mazzanti A, Svelto F (2015) A 40-67 GHz power amplifier with 13 dBm PSAT and 16% PAE in 28 nm CMOS LP. IEEE J Solid-State Circuits 50(7):1618–1628
4. Fu Y, Li L, Wang D, Wang X (2020) A -193.6 dBc/Hz FoMT 28.6-to-36.2 GHz Dualcore CMOS VCO for 5G applications. IEEE Access 8:1–1
5. Gupta A, Member S, Jha RK, Member S (2015) A survey of 5G network: architecture and emerging technologies. IEEE Access 3:1206–1232
6. Li CH, Kuo CN, Kuo MC (2012) A 1.2-V 5.2-mW 20-30-GHz wideband receiver front-end in 0.18-μm CMOS. IEEE Trans Microw Theory Technol 60(11):3502–3512
7. Li YA, Hung MH, Huang SJ, Lee J (2010) A fully integrated 77 GHz FMCW radar system in 65 nm CMOS. Digest in Technical papers—IEEE international solid-state circuits conference, vol 53, pp 216–217
8. Kuo KC, Leuciuc A (2001) A linear MOS transconductor using source degeneration and adaptive biasing. IEEE Trans Circuits Syst II Analog Digit Signal Process 48(10):937–943
9. Božanić M, Sinha S (2020) Millimeter-wave integrated circuits: methodologies for research, design and innovation, vol 658. Springer
10. Bronckers LA, Member S, Roc A, Smolders AB, Member S (2020) Wireless receiver architectures towards 5G : Where are we ?," vol 17
11. Oh NJ, Lee SG (2005) Building a 2.4-GHz radio transceiver using IEEE 802.15.4. IEEE Circuits Devices Mag 21(6):43–51
12. Vitee N, Ramiah H, Chong WK (2015) A wideband CMOS LNA-mixer for cognitive radio receiver. In: IEEE Asia-Pacific conference on circuits systems proceedings, APCCAS, vol 2015–February, pp 348–351
13. Wu C, Lin Y, You H, Huang S (2014) A 1.04 mm2, low-power front-end circuit of receiver in 0.18 μm CMOS for IEEE 802.11a application, pp 427–430
14. Chironi V, Pasca M, Siciliano P, D'Amico S (2015) A 5.8–13 GHz SDR RF front-end for wireless sensors network robust to out-of-band interferers in 65 nm CMOS. In: Proceedings—2015 6th IEEE International workshop in advances sensors interfaces, IWASI 2015, pp 137–140
15. El-Desouki MM, Qasim SM, BenSaleh MS, Jamal Deen M (2015) Toward realization of 2.4 GHz balunless narrowband receiver front-end for short range wireless applications. Sensors (Switzerland) 15(5):10791–10805
16. Hqvkhq LL, Dqg DQJ, Kdr X, Dqg E, Kljkhu D, Yld D, Dojrulwkpv V, Zlwk VV, Xvhuv HQG, Uhtxluhphqwv WVV, Suhvhqwv S, Hiilflhqw DQ, Ri G, Wudqvplwwhu QWKH, Lj P, Wkh S, Zdyhirup P (2018) A 5.8 GHz DSRC digitally controlled CMOS RF = SoC Transceiver for China ETC, pp 323–324
17. Cao L, Liu R, Zhang Y, Zhang M, Li X, Wang W, Liu H, Lu C (2018) A 5.8 GHz digitally controlled CMOS receiver with a wide dynamic range for Chinese ETC system. IEEE Trans Circuits Syst II Express Briefs 65(6):754–758
18. Chong WK, Ramiah H, Tan GH, Vitee N, Kanesan J (2014) Design of ultra-low voltage integrated CMOS based LNA and mixer for ZigBee application. AEU—Int J Electron Commun 68(2):138–142

19. Li Z, Cheng G, Wang Z (2018) A 0.1–1 GHz low power RF receiver front-end with noise cancellation technique for WSN applications. AEU—Int J Electron Commun 83:288–294
20. Vigilante M, Member S, Reynaert P, Member S (2017) On the design of wideband transformer-based fourth order matching networks for E -band receivers in 28-nm CMOS, pp 1–12
21. Bhat AN, Van Der Zee R, Finocchiaro S, Dantoni F, Nauta B (2019) A baseband-matching-resistor noise-canceling receiver architecture to increase In-band linearity achieving 175 MHz TIA bandwidth with a 3-Stage Inverter-Only OpAmp," pp 1–4
22. Nam H, Member S, Lee W, Son J, Park J, Member S, Abstract A, IQ (2020) A compact I/Q upconversion chain for a 5G wireless transmitter in 65-nm. CMOS Technol 30(3):284–287
23. Medra A, Giannini V, Guermandi D, Wambacq P (2014) A 79 GHz variable gain low-noise amplifier and power amplifier in 28 nm CMOS operating up to 125°C. Eur Solid-State Circuits Conf 2(1):183–186
24. Mineyama K, Akiko K, Yoichi S, Masaru S, Toshihide H, Naoki J (2011)A millimeter-wave CMOS low noise amplifier using transformer neutralization techniques. In: Proceedings Asia–Pacific microwave conference, pp 223–226
25. Wu C, Member S, Wang Y, Nikolić B, Member S, Hull C, Member S (2016) An interference-resilient wideband mixer-first correlated orthogonal calibration. Trans Microw Techiques 64(4):1088–1101
26. Murphy D, Darabi H, Abidi A, Hafez AA, Mirzaei A, Mikhemar M, Chang MCF (2012) A blocker-tolerant, noise-cancelling receiver suitable for wideband wireless applications. IEEE J Solid-State Circuits 47(12):2943–2963
27. Lien YC, Klumperink EAM, Tenbroek B, Strange J, Nauta B (2018) Enhanced-selectivity high-linearity low-noise mixer-first receiver with complex pole pair due to capacitive positive feedback. IEEE J Solid-State Circuits 53(5):1348–1360
28. Banerjee A (2016)A 45 nm CMOS RF-to-Bits LTE/WCDMA FDD/TDD 2 × 2 MIMO base-station transceiver SoC with 200 MHz RF bandwidth. In: IEEE international solid-state circuits conference, pp 164–166
29. McLaurin DJ, Gard KG, Schubert RP, Manglani MJ, Zhu H, Alldred D, Li Z, Bal SR, Fan J, Gysel OE, Mayer CM, Montalvo T (2018) A highly reconfigurable 65 nm CMOS RF-to-bits transceiver for full-band multicarrier TDD/FDD 2G/3G/4G/5G macro basestations. In: Digest in technical papers - IEEE international solid-state circuits conference, vol 61, pp 162–164
30. Lee JG, Lee HJ, Kim SH, Byeon CW, Park CS (2017) 60 GHz direct up-conversion mixer with wide if bandwidth and high linearity in 65 nm CMOS. In: 2017 IEEE International symposium on radio-frequency integrated technology RFIT 2017, pp 74–76
31. Byeon CW, Lee JH, Lee DY, Kim MR, Son JH (2015) A high linearity, image/LO-rejection I/Q up-conversion mixer for 5G cellular communications. Eur Microw Week 2015:345–348
32. Won Y, Kim C, Lee S (2015) A 24 GHz highly linear up-conversion technology, pp 22–24

LC VCOs for 5G Networks

Sankaran Aniruddhan

Abstract This chapter deals with LC VCOs for 5G applications. The operating principle of LC VCOs based on the negative resistance model is first introduced. This leads to a discussion of phase noise and its standardized analytical form, including a primer on various phase noise models. Design considerations and challenges for LC-VCOs at 5G mm-wave frequencies are then described. The chapter concludes with a description of circuit analysis, design techniques and implementation of two millimeter wave VCOs recently reported in the literature.

Keywords Bipolar oscillator · Phase noise · Phase delay analysis · Class-C · Discrete tuning range · Continuous tuning range

1 Introduction

There has been a sustained effort over the last few decades to integrate the RF front-end of transceivers with the digital baseband portion to develop low-cost, low-power and smaller wireless products that deliver increasingly higher data rates. CMOS has emerged as the technology of choice for these System On-Chip applications due to its high integrability and relative low cost as compared with other technologies like Bipolar and GaAs. Voltage Controlled Oscillators (VCOs) are key blocks in the transceiver Local Oscillator (LO) path that enable frequency translation and allow channel selection of the RF signals. Full integration of CMOS VCOs presents a continuing challenge for low cost high-performance 5G transceivers. In particular, as frequencies increase, there are various layout related issues that crop up. Similarly, as the distance between various subblocks of the oscillator becomes a significant fraction of the wavelength, delays inside the circuit can no longer be ignored. These become more important in circuits like millimeter wave oscillators which employ feedback. Finally, the quality of on-chip passive components—spiral inductors, MIM

S. Aniruddhan (✉)
Department of Electrical Engineering, Indian Institute of Technology Madras, Chennai, India
e-mail: ani@ee.iitm.ac.in

S. Singh et al. (eds.), *CMOS Analog IC Design for 5G and Beyond*, Lecture Notes in Electrical Engineering 719, https://doi.org/10.1007/978-981-15-9865-4_6

capacitors and varactors—remains a critical bottleneck in integrated high-frequency VCO design.

2 LC Oscillators

LC VCOs normally produce a much more spectrally pure signal compared with most other types of oscillators. In differential configurations, they also offer good rejection of common-mode noise from the power supply and ground. The LC VCO can be viewed as a closed-loop feedback system with a forward function consisting of an amplifier, and a feedback function consisting of a frequency selective network, which is usually a parallel LC (inductor-capacitor) tank, as shown in Fig. 1. Barkhausen's criteria for feedback systems set two simultaneous conditions for oscillation: magnitude of loop gain = 1 and phase shift around the loop = 360°. This view is useful in describing the operation of single-ended oscillator configurations such as the Colpitts, Hartley and Pierce oscillators [1].

However, modern integrated LC VCOs are better viewed as a parallel combination of an active circuit and a parallel LC resonator, as shown in Fig. 2. A nonideal LC tank has finite Q and presents equivalent resistive impedance at resonance. The active circuit can hence be viewed as providing an equivalent negative resistance to cancel out the positive resistance of the tank at resonance, and hence sustaining the oscillations. From energy considerations, we can say that the active circuit adds just enough energy to replenish the total losses in the tank circuit over one oscillation

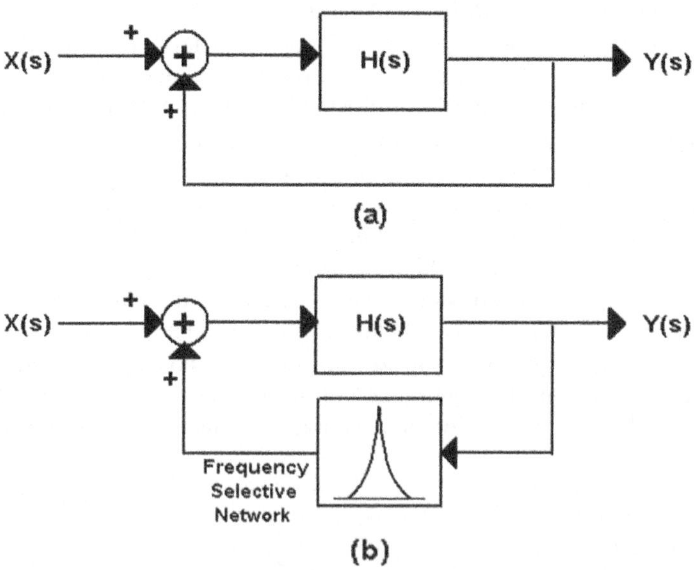

Fig. 1 **a** Feedback oscillatory system, **b** frequency-selective network added to (**a**)

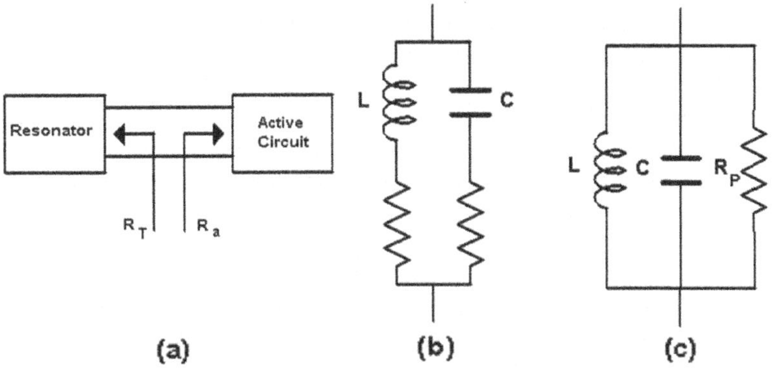

Fig. 2 (**a**) One port view of oscillators, (**b**) LC resonator, (**c**) Equivalent circuit of (**b**)

cycle. If the negative resistance of the active circuit is not sufficient to compensate for the losses in the LC tank, the oscillations never start, or if already started, die down quickly. There are several ways of combining the resonator with transistors to create an oscillator, leading to many possible oscillator architectures. Traditional ones such as Colpitts, Hartley and Pierce topologies implement feedback from tapped inductor/capacitor dividers, with a single transistor providing the active gain. The transistor is variously connected in common-source or common-gate configurations. In recent times, however, the so-called—gm or cross-coupled oscillator has become extremely popular in integrated LC oscillators due to its simple configuration and convenient IC realization. Differential circuit topologies are generally preferred in integrated circuits due to their better rejection of supply noise. They also perform better in the presence of substrate-coupled switching noise from digital circuits on the same chip. Finally, most active and passive mixers are designed in differential configuration, requiring differential LO drives. The gm oscillator has an intrinsic differential topology, eliminating the need for any single-ended to differential conversion circuitry.

2.1 *Negative gm VCO Design*

The design of the gm VCO begins with the design of the resonator, which is made up of an inductor and capacitor in parallel. The inductor is usually implemented as an on-chip spiral inductor to integrate it with the rest of the VCO circuit. The resonator capacitance consists of fixed and variable components. The fixed component usually consists of the parasitic capacitance of the active devices and the fixed capacitance of the varactor, while the variable component is made up of the adjustable capacitance of the varactor. The varactor is realized as either a reverse biased junction diode or a MOS device in accumulation or inversion modes.

The tank impedance characteristics are plotted over frequency and the equivalent tank resistance at the required oscillation frequency, R_P, is obtained. The cross-coupled MOSFET pair is designed for a net negative resistance that is smaller than R_P for sustained oscillations (note that since the impedances appear in parallel, the negative resistance should be smaller than the positive resistance to give a net negative resistance). It is essential that the VCO be designed with a net small-signal conductance that is negative, so that the small signal loop gain is greater than unity (say around 2 to 3) for quick startup of the oscillation waveform. It is easily shown from Fig. 3 that the small-signal input impedance of a simple cross-coupled NMOS pair is –2/gm. Therefore, the gm of the transistors is decided based on the tank parallel resistance along with the required design margin (say, $-R_P/2$) at the oscillation frequency, f_0. At resonance, the tank inductor-capacitor parallel combination offers a purely real impedance R_P across its terminals. Once the tank parallel resistance is known, the required output amplitude sets the bias current value. The transistor sizes (W/L) are fixed once the gm and bias current are decided. Figure 4 shows the circuit topology of a cross-coupled LC-tuned VCO. Let C_P be the fixed capacitance at each output node due to the gate and drain parasitic capacitance of each MOSFETs. The frequency of oscillation of the VCO is given by

$$f_0 = \frac{1}{2\pi\sqrt{\frac{L}{2}(2C + C_P)}} = \frac{1}{2\pi\sqrt{L(C + \frac{C_P}{2})}} \tag{1}$$

The cross-coupled MOSFET pair forms a positive feedback combination, which produces negative resistance, and behaves, as a switching current source. Due to the positive feedback, we may assume that at any point of time, only one of the MOSFETs is fully on while the other is fully off. The current through each MOSFET switches between I_{bias} and zero. The AC equivalent half-circuit of the VCO is illustrated in

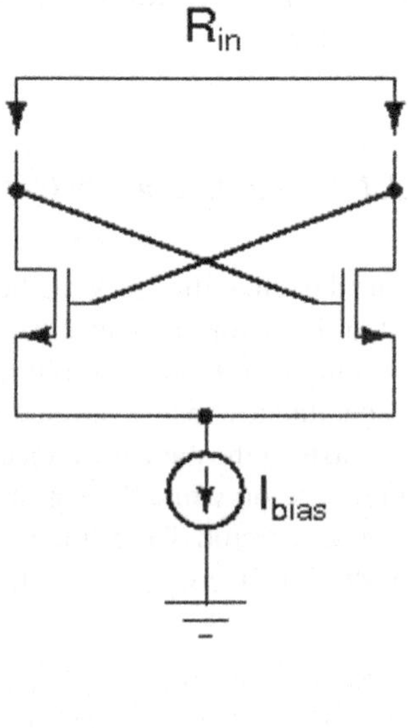

Fig. 3 Input impedance of cross-coupled NMOS pair

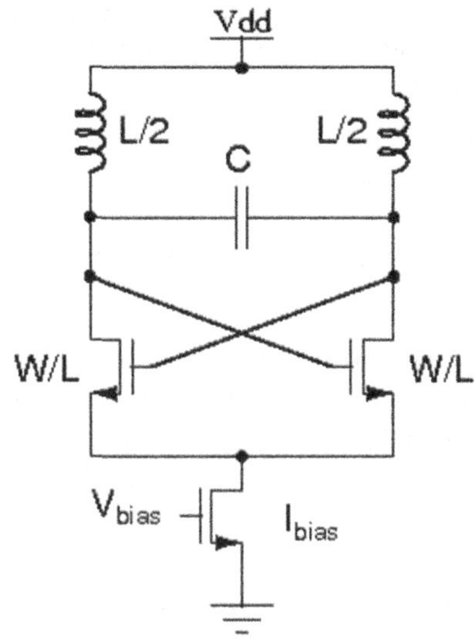

Fig. 4 NMOS cross-coupled LC VCO

Fig. 5 [2]. The LC tank filters this square wave signal and the main component of the output is the fundamental tone, f0, which is also the desired oscillation frequency. Hence the amplitude of the voltage across the tank half-circuit is

$$V_{\tan k,se} = \frac{1}{\pi} I_{bias} R_P \tag{2}$$

and the differential output voltage across the complete LC tank is

$$V_{\tan k,diff} = \frac{2}{\pi} I_{bias} R_P \tag{3}$$

The complementary version of the LC VCO shown in Fig. 4 has both NMOS and PMOS cross-coupled devices and employs current reuse to generate twice the negative resistance at the same power consumption. Current flows through the inductor in

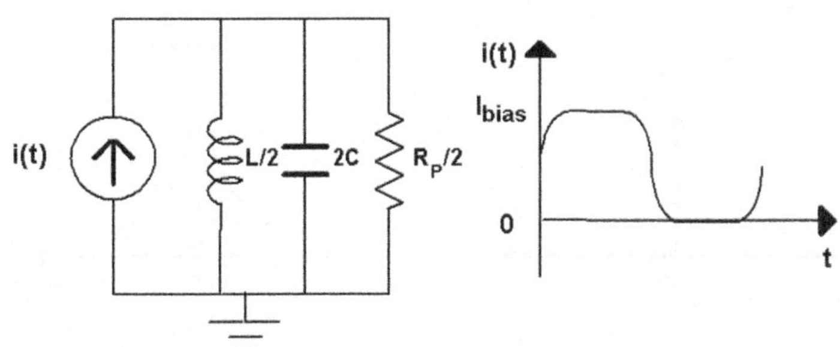

Fig. 5 Switching I_{bias} for an NMOS cross-coupled VCO [2]

both half-cycles, as opposed to the all-NMOS VCO, where current flows through each half-inductor only for one half-cycle. Thus the CMOS VCO has twice the voltage amplitude as an all-NMOS VCO for the same bias current, and the fundamental of the switching current is $4/\pi$ times the amplitude of the square wave.

2.2 *Phase Noise*

"Phase noise" in general refers to the random phase and frequency fluctuations of a signal. The output of an ideal VCO is a pure sinusoid, but the output of a real oscillator can be represented by

$$V_{out} = A(t)\cos(\omega_o t + \phi(t)) \tag{4}$$

where A(t) is the noise in the amplitude and φ(t) is the phase variation in the output of the VCO due to noise. Due to this phase noise, the effective frequency of the oscillator is given by

$$\omega = \omega_o + \frac{d\varphi(t)}{dt} \tag{5}$$

This effect is called "Jitter" when viewed in the time-domain as a random change in the period (or zero-crossings) of the signal. The output spectra of an ideal and actual VCO are shown in Fig. 6 [1]. The ideal oscillator has an output, which is an impulse at ω0, while an actual oscillator has noise "sidebands" around ω_0. The phase noise at a frequency offset $\Delta\omega$ from the center frequency is defined as the ratio of noise power in a 1 Hz bandwidth at the $\Delta\omega$ offset, to the carrier power at the fundamental frequency of oscillation ω0, and is measured in units of dBc/Hz. Therefore, phase noise is inversely proportional to the square of the oscillation voltage amplitude. Due to the switching behavior of the cross-coupled transistor pair in the VCO, frequency translation of noise occurs and low-frequency noise is modulated or up-converted to the oscillation frequency ω_0. Phase noise is a major concern in

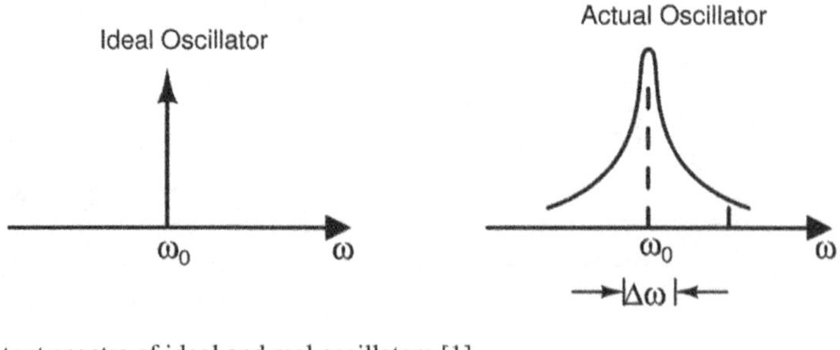

Fig. 6 Output spectra of ideal and real oscillators [1]

present day communication systems due to effects such as reciprocal mixing and transmit-receive leakage [1]. Since phase noise is measured relative to the carrier amplitude and decreases as the square of voltage amplitude, the phase noise for a complementary cross-coupled VCO should ideally be 6 dB lower than that for an NMOS-only cross-coupled VCO. But the presence of the extra PMOS transistors doubles the number of active-device noise sources, thereby degrading the phase noise by around 3 dB. The overall phase noise of a CMOS VCO is therefore better than an all-NMOS VCO by about 3 dB. The two cross-coupled pairs, being in parallel, generate twice the negative transconductance for the same current. The NMOS and PMOS devices are sized so as to generate approximately equal negative resistances. This results in symmetric rise and fall times for the sinusoidal voltage at the tank nodes, leading to a lower up-conversion of low-frequency noise to f0 [3]. Phase noise and associated mechanisms is discussed in more detail later in the chapter. The major drawback of the complementary VCO is the presence of PMOS devices which, being weaker than NMOS devices, need to be sized larger than the NMOS devices. This leads to a larger parasitic capacitance across the tank and hence reduction in tuning range. At mm-wave frequencies, this additional parasitic capacitance is usually unacceptable, and NMOS-only VCOs are preferred in most high-frequency applications.

2.3 LC VCO Tuning

Since the LC product fixes the oscillation frequency of the VCO, frequency tuning is achieved by controlling the tank capacitance with a DC voltage. Reverse-biased p-n junctions exhibit a change in capacitance with voltage, by modulation of the depletion region width with the applied voltage. Such devices that are designed specifically for voltage-variable capacitance are called Varactors. The capacitance of a reverse-biased junction diode varies with the applied voltage according to the following equation

$$C_j = \frac{C_{jo}}{\left(1 - \frac{V_F}{\phi}\right)^n} \tag{6}$$

where n = 1/2 for abruptly doped junctions. This can cause a somewhat nonlinear variation of capacitance with applied voltage. Linearly graded junctions have n = 1/3, thereby making their voltage-to-capacitance transfer characteristic more linear compared with the previous case. VCO frequency fine-tuning is accomplished using varactors, since they can be designed to be quite linear over small changes in voltage, and also because they offer a continuous variation in capacitance. Some form or other of digital tuning is employed quite commonly for coarse tuning of oscillators. One or more capacitors from a bank of capacitors are switched in parallel with the VCO tank, thereby changing the center frequency by a discrete value. The switches are implemented as series MOSFET devices. When switched on, the series resistance of these MOS switches can degrade the quality factor of the LC tank, and hence the phase noise performance of the VCO. Though the Q of the tank capacitance is higher than that of the tank inductance, the overall Q is considerably affected by any change in either Q. The condition for minimum Q of the capacitor bank corresponds to the case of maximum resistive loss, and occurs when all the series switches are on, i.e. at the minimum frequency of oscillation of the VCO. Consequently, the worst-case phase noise could potentially occur at this end of the tuning range, which is often unexpected. Under these conditions, the worst-case Q of the capacitor bank is given by ω0CRP C where C is the total capacitance of the bank and RP C is the equivalent parallel resistance across the bank due to the series resistance of the switches. RF junction varactors and MOS varactors usually have a higher Q than these capacitor banks.

2.4 LC VCO Optimization

So far, we have seen basic aspects of VCO design, such as device sizing (based on negative resistance required for startup), bias current (based on power consumption and output amplitude) and LC product (based on frequency of oscillation). The basic equations governing the output amplitude of the VCO were also discussed. The output voltage amplitude across the tank has a significant effect on phase noise, since phase noise is always expressed relative to the amplitude of the fundamental frequency of oscillation. There are two regimes of operation of gm VCOs: Current-limited and voltage-limited regimes [2]. In the current-limited regime, the output voltage amplitude is determined solely by the bias current value and the equivalent tank resistance at resonance. In this region of operation, increasing the bias current increases the output amplitude and hence improves the phase noise performance of the VCO. However, the output voltage amplitude cannot be increased indefinitely. Eventually, when the voltage amplitude across the tank approaches the supply voltage value, the cross-coupled devices the tail bias current source will spend increasing amounts of time in the linear region of operation, especially at the peaks of the output sinusoidal voltage waveforms. Any further increase of bias current will not increase the output voltage amplitude significantly, and the VCO is said to operate in the Voltage-limited regime of operation.

Figure 7 shows the tank voltage amplitude as a function of bias current and the two regimes of operation for a gm VCO. In the voltage-limited region, any increase in bias current does not increase carrier power, and the additional power dissipation goes only toward increasing noise power, and phase noise performance degrades. On the other hand, in the current-limited region, any increase in the bias current raises the carrier power faster than the noise power, and hence the phase noise performance improves until the voltage-limited regime is approached. Best VCO phase noise performance is usually achieved near the transition region between the two regimes

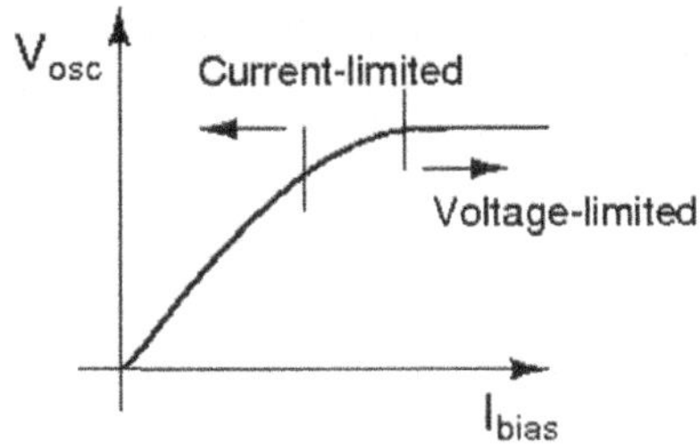

Fig. 7 Current-limited and voltage-limited regimes of operation of a −gm VCO

of operation. If power consumption is not a concern at all in the design of the VCO, the all-NMOS VCO may be the preferred topology because it can have a higher voltage swing and can therefore be designed to have a better phase noise performance than the complementary VCO.

2.5 LC VCO Bias Circuit

Design of the bias circuit of the VCO is quite critical to the performance of the VCO. It is known from phase noise theory that thermal and flicker noise from the bias device are upconverted to f0, thereby degrading the phase noise [3]. The bias is usually designed as a current mirror and the necessary amount of bias current is mirrored into the VCO core. This is also quite convenient since the performance of the VCO with change in bias current can be easily measured. The value of the bias current (also called the tail current) sets the power consumed in the oscillator. The voltage-headroom available for the current source is (V_{DD}—V_{GSN}—V_{GSP}), which depends on the bias current, supply voltage and the sizes of the cross-coupled transistors, but is typically around 0.2 V. In most cases, cascode-type bias circuits are precluded, and simple current mirror circuits are the common option. PMOS transistor bias may not be an option since the size required for such a low V_{DSAT} may be too large. On the other hand, PMOS transistors may offer less noisy bias circuits. It is also possible to build a voltage-biased VCO by using a low-dropout regulator (LDO) to bias the common node of the inductor halves (also called the center-tap of the complete inductor). It is not advisable to completely avoid any bias circuit, as the VCO performance will become a strong function of process, voltage and temperature.

2.6 Phase Noise in LC VCOs

Phase noise exists in an autonomous oscillator because it has no time reference, and no phase limiting mechanism. Therefore, its phase is free to drift, and any noise that induces a time-shift causes a phase change that cannot be restored by the oscillator. In accordance with the equipartition theory of thermodynamics, this noise exhibits itself equally as amplitude and phase noise [3]. Most LC oscillators have an inherent nonlinear amplitude stabilization mechanism that suppresses the amplitude variations and the amplitude noise thus remains small. Consequently, the major portion of the noise in an oscillator exhibits itself as phase noise, causing time-dependent variations in the phase of the oscillator. This results in noise sidebands around the frequency of oscillation in the output spectrum of the oscillator. Moreover, some of the AM noise that is present in the oscillator is converted to phase noise through the varactors present in the circuit. For these reasons, only phase noise is considered in the design of RF LC VCOs. In the frequency spectrum of the output of the oscillator, this phase noise causes high levels of noise near the carrier frequency f_0. Since the noise is predominantly in the phase of the oscillator, it cannot be eliminated by passing the signal through a limiter as in the case of amplitude noise. Since the noise appears very close to the carrier frequency as sidebands, it cannot be eliminated by filtering. The chief phase noise sources in the VCO circuit are the flicker and thermal noise of the cross-coupled devices and the bias devices, as well as the thermal noise from the LC-tank. The thermal noise from the tank originates mainly from the tank resistance due to the finite Q of the inductor and varactor. The band-pass frequency response of the LC-tank shapes the thermal noise from the devices and the tank itself. In addition, if some sort of digital tuning scheme is used, there can be a significant contribution of flicker and thermal noise from the switch devices in the capacitor bank. The frequency response of the LC tank affects the noise contribution from different sources in the circuit. At the oscillation frequency, which is the same as the resonant frequency of the tank, its impedance is purely resistive. The impedance of the tank at a frequency offset $\Delta\omega$ from the oscillation frequency $\omega_0 = 2\pi f_0$ can be expressed in terms of the tank parameters as [4].

$$z_T(\omega_o + \Delta\omega) = j\frac{\omega_o L}{2\frac{\Delta\omega}{\omega_o}} \tag{7}$$

which can be simplified to

$$|z_T(\omega_o + \Delta\omega)| = R_P \frac{\omega_o}{2Q\Delta\omega} \tag{8}$$

The expression for the equivalent mean-square thermal noise current source for a tank with equivalent parallel resistance R_P is

$$\overline{i_n^2} = \frac{4kT}{R_P}\Delta\omega \tag{9}$$

Figure 8 shows the equivalent noise current source across the tank. Assuming Linear Time Invariant (LTI) behavior, the resulting voltage noise power spectral density can be determined as

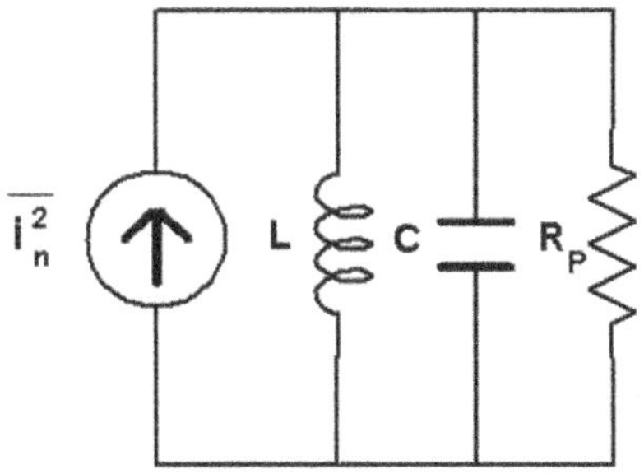

Fig. 8 Thermal noise shaped by frequency response of tank

$$\frac{\overline{v_n^2}}{\Delta\omega} = \frac{\overline{i_n^2}}{\Delta\omega}|z_T(\omega_o + \Delta\omega)|^2 \tag{10}$$

which can be further simplified to

$$\frac{\overline{v_n^2}}{\Delta\omega} = 4kTR_P \cdot \left(\frac{\omega_o}{2Q\Delta\omega}\right)^2 \tag{11}$$

Therefore, the voltage phase noise due to thermal noise is frequency dependent due to the filtering action of the tank and has a $1/\omega^2$ variation around the center frequency ω_0. The above expression for the noise voltage conveys some valuable information regarding the phase noise. First, phase noise can be minimized by reducing the number of noise sources in the system. Since active devices are significant sources of noise, the number of transistors in the oscillator should be minimized. Second, the frequency response of the tank should be made as narrow as possible. This can be accomplished by implementing the VCO with a high-Q LC tank. Furthermore, the phase noise is inversely proportional to the square of the quality factor of the tank [5]. Consequently, increase in tank Q allows us to decrease phase noise at the same VCO power dissipation or operate the VCO at a lower power for the same phase noise specification.

The Single Side-band (SSB) phase noise of an oscillator, at a frequency offset Δf from the fundamental frequency of oscillation f_0 is defined as

$$L\{\Delta\omega\} = 10\log_{10}\left(\frac{P_{1Hz}(\omega + \Delta\omega)}{P_S}\right) \tag{12}$$

where P_{1Hz} is the noise power in a 1 Hz bandwidth at the Δf offset and P_S is the signal power at f_0. Since it is calculated relative to the carrier power, it is measured in units of dBc/Hz where "c" stands for "carrier." The phase noise variation with Δf is represented in a log–log plot and illustrated in Fig. 9.

Three distinct regions can be identified in this plot [5]. First, a $1/f^2$ region due to filtering of thermal noise of the tank and active devices by the tank impedance frequency response. This region of the phase noise plot has a slope of −2 on the log–log plot. Next, a $1/f^3$ region exists due to up-converted flicker noise from the bias

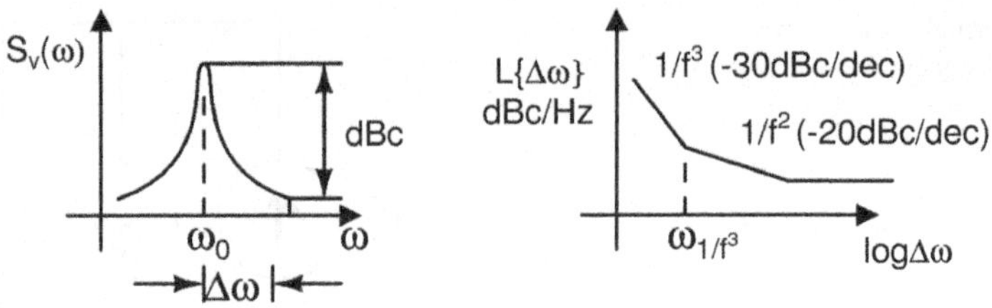

Fig. 9 General SSB phase noise plot

and cross-coupled devices, which is superimposed onto the $1/f^2$ region. The phase noise in this region has a slope of −3 in a log–log plot. This is dominant at lower frequency offsets, and the offset frequency at which the phase noise transition from $1/f^3$ to $1/f^2$ behavior is called the 1/f corner frequency (f_1/f^3). This is the frequency offset at which the slope of phase noise changes from −3 to −2. Finally, the plot contains a flat, white-noise region at larger frequency offsets due to the noise floor of the buffers and the measurement equipment, with zero slope in the log–log plot.

2.7 Leeson's Phase Noise Model

Leeson's model for the phase noise of an oscillator has been quite popular, and is based on Linear Time-Invariant (LTI) analysis of the LC oscillator in steady state using feedback theory [5]. According to Leeson's model, the feedback oscillator noise spectrum is given by

$$L\{\Delta\omega\} = 10\log_{10}\left(\frac{2FkT}{P_S}\left(1+\frac{\omega_o}{2Q\Delta\omega}\right)^2\cdot\left(1+\frac{\omega_{1/f^3}}{|\Delta\omega|}\right)\right) \tag{13}$$

where F is an empirically determined parameter called the "device excess noise number," k is Boltzmann Constant, T is absolute temperature in K and $\omega^3_{1/f} = 2\pi f_1/f^3$. The factor F accounts for measured excess phase noise over the calculated expression in the $1/f^2$ region, while the factor of unity in the braces accounts for the noise floor observed at higher frequency offsets. The factor in the second set of braces is brought into provide a 1/f 3 behavior at low frequency offsets. Since F is determined from actual measurements, the model does not actually predict the exact phase noise expected in a VCO. Moreover, the 1/f phase noise corner frequency, f_1/f^3, is assumed to be the same as the 1/f corner frequency of device noise. Measured corner frequencies, however, often do not match the device corner frequency, and the model does not account for multiple sources of 1/f noise in the VCO.

2.8 The Rael-Abidi Phase Noise Model

The second phase noise model we discuss in this chapter is the Rael-Abidi model [6, 7]. The fundamental result of this work is the calculation of a precise expression for the empirical factor F in Leeson's model by identifying the various nonlinear physical processes by which noise from different parts of the circuit get converted to phase noise around f_0 in the $1/f^2$ region. Leeson's model for the phase noise of a differential LC oscillator is also validated by this theory. The theory is developed for an all NMOS oscillator but can also be extended to a Complementary VCO. We saw earlier that the cross-coupled transistors in the −gm oscillator can be considered as a switching current source of amplitude I_{bias} across the LC tank, where I_{bias} is the tail bias current. The sinusoidal voltage across the resonator drives the differential pair into switching, thus sustaining the oscillations. We will first assume that the differential pair is ideal, and a small voltage excursion around the zero crossing of the voltage across the resonator is sufficient to switch the transistor pair. Consider a noise voltage $v_n(t)$ superposed on top of the output sinusoidal waveform. This noise voltage changes the instantaneous value of the sinusoidal voltage by a small amount and hence advances or retards the instant at which the differential pair switches. The randomly pulse-width modulated output current can then be decomposed into the sum of an ideal square-wave of amplitude I_{bias} and a series of pulses of constant height but random width. The pulses can further be approximated by the product of the original $v_n(t)$ waveform and an impulse train at twice the oscillation frequency, corresponding to an impulse at every zero crossing. Next, noise from different sources in the VCO—Resonator Noise, Tail Current Noise and Cross-coupled Pair Noise—are considered in turn to determine the contribution to phase noise from each source. The total phase noise in the $1/f^2$ region is obtained by summing the expressions for phase noise due to all the above sources and is compared to Leeson's phase noise model in Eq. (13). This gives the following expression for F of a differential cross-coupled LC oscillator

$$F = 2 + \frac{8\gamma R_P I_{bias}}{\pi V_o} + \gamma \frac{8}{9} g_{m,bias} R_P \tag{14}$$

where V_0 is the output voltage amplitude of the oscillator, $g_{m,bias}$ is the transconductance of the bias device, and γ is the excess noise factor of the MOSFETs. Thus Leeson's phase model is validated, and all nonlinear and frequency-mixing effects are accounted for. Flicker noise in the tail current source is hypothesized to get upconverted through the mixer-action of the differential pair, but this appears as AM noise. Unless a high-gain varactor is present to convert this from AM to FM noise, this does not appear as phase noise. The flicker noise in the differential pair is sampled by the impulse train with frequency equal to twice the oscillation frequency, and therefore appears as phase noise around DC and twice the oscillation frequency. Since both these frequencies are filtered due to the tank impedance frequency response, this mechanism does not explain $1/f^3$ close-in phase noise. Therefore, the above model does not fully account for flicker noise up-conversion from bias and cross-coupled transistors.

2.9 The Hajimiri-Lee Phase Noise Model

The third and final model discussed here is the Hajimiri-Lee Phase Noise Model, which is a more accurate model for the phase noise of a VCO [3, 4]. The assumptions of linearity and time-invariance applied in deriving the LTI Leeson's phase noise model are examined and revised. It is first shown that the assumption of linearity is not entirely valid because nonlinearity is a basic property of all practical oscillators. In real VCO, amplitude stabilization is achieved using some kind of nonlinear mechanism. However, linearization can be applied around the large signal steady state of the oscillator, which implicitly takes into account the nonlinearity of the oscillator. Since the noise power is several orders of magnitude smaller than the signal power, linearity is an acceptable assumption for the noise-to-phase transfer function. However, it is quite simple to show that even ideal oscillators are time-variant systems. Consider an ideal LC tank excited by a current pulse as shown in Fig. 10. Before application of the current pulse, assume that the system is oscillating with some steady-state amplitude. We can analyze the response of the system when the current impulse is applied at different time instants. Noise charge injected into the LC tank is represented as a current impulse, which flows completely into the capacitance, and results in an instantaneous voltage change across the capacitor. It can be easily seen from Fig. 11 that the resultant change in the amplitude and phase of the oscillator is time dependent. When the impulse is applied at the peak of the oscillation amplitude, only the amplitude of the waveform is changed, while the phase is unaffected. On the other hand, if the impulse is applied at the zero crossing of the voltage waveform, the amplitude remains unaltered, while having maximum effect on the phase of the signal. An impulse applied between these two points of time results in both amplitude and phase change. Thus, the response of an LC tank to any noise is inherently time-dependent. The intrinsic nonlinearity of a practical oscillator quenches the amplitude noise over time, while any phase fluctuation persists indefinitely. Therefore, the impulse response of an oscillator is a time-varying unit step. Since the linearity assumption is valid for the current-to-phase transfer function, the oscillator is in some sense a Linear Time-Variant (LTV) system for noise purposes.

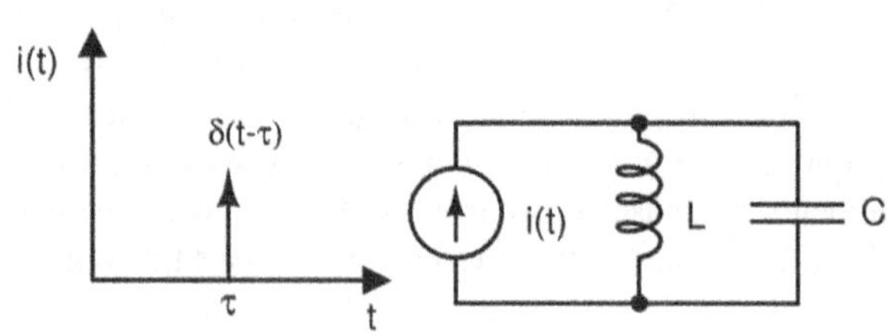

Fig. 10 LC oscillator excited by a current pulse

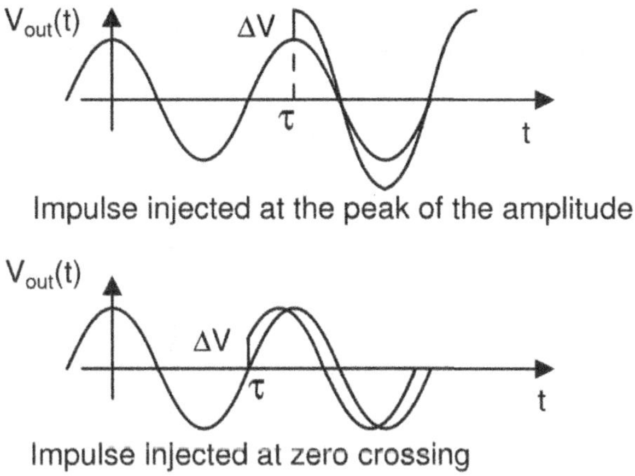

Fig. 11 LC oscillator excited by a current pulse

We can express the excess phase φ(t) to be proportional to the ratio of injected charge to the maximum charge swing across the capacitor, and determine the unit impulse response for excess phase as completely into the capacitance and results in an instantaneous voltage change across the capacitor. It can be easily seen from Fig. 11 that the resultant change in the amplitude and phase of the oscillator is time dependent. When the impulse is applied at the peak of the oscillation amplitude, only the amplitude of the waveform is changed, while the phase is unaffected. On the other hand, if the impulse is applied at the zero crossing of the voltage waveform, the amplitude remains unaltered, while having maximum effect on the phase of the signal. An impulse applied between these two points of time results in both amplitude and phase change. Thus, the response of an LC tank to any noise is inherently time dependent. The intrinsic nonlinearity of a practical oscillator quenches the amplitude noise over time, while any phase fluctuation persists indefinitely. Therefore, the impulse response of an oscillator is a time-varying unit step. Since the linearity assumption is valid for the current-to-phase transfer function, the oscillator is in some sense a Linear Time-Variant (LTV) system for noise purposes. We can express the excess phase φ(t) to be proportional to the ratio of injected charge to the maximum charge swing across the capacitor, and determine the unit impulse response for excess phase as

$$h_{\phi}(t,\tau) = \frac{\Gamma(\omega_o t)}{q_{\max}} u(t-\tau) \tag{15}$$

where q_{max} is the maximum charge displacement across the capacitor and u(t) is the unit step. Γ (x) is a dimensionless, frequency- and amplitude-independent, periodic function with period 2π, called the Impulse Sensitivity Function (ISF). The ISF describes the sensitivity of an oscillator to an impulse applied at phase ω0t. For example, in an LC oscillator, the ISF has its maximum value near the zero crossings and zero value at the maxima of the sinusoidal oscillation waveform. The ISF can

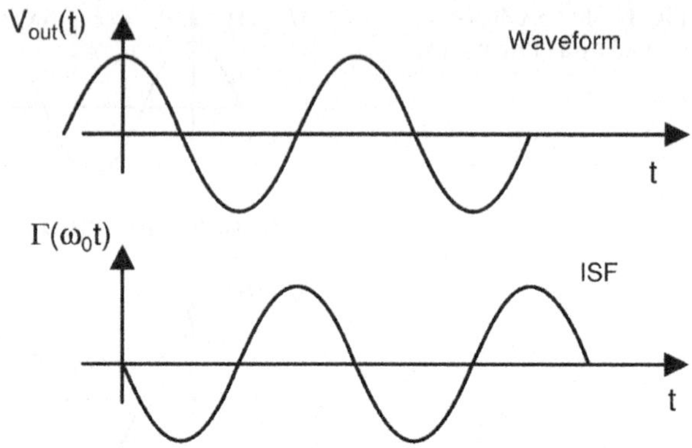

Fig. 12 Sinusoidal waveform and corresponding ISF of an LC oscillator

be obtained most accurately through simulation, but approximate analytical expressions can also be utilized. If the oscillation waveform function is f, an approximate expression for ISF is

$$\Gamma(x) = \frac{f'}{(f')^2 + (f'')^2} \tag{16}$$

where f' and f'' are the first and second derivatives of f respectively. For ideal sinusoidal oscillations, f = cos(x), so Γ $(\omega_0 t)$ = -sin$(\omega_0 t)$. The ISF of an LC oscillator is shown in Fig. 12.

Since the ISF is periodic, it can be expanded into a Fourier series as

$$\Gamma(\omega_o t) = \frac{c_o}{2} + \sum_{n=1}^{\infty} c_n \cos(n\omega_o t + \theta_n) \tag{17}$$

where the Fourier series coefficients cn are real valued, and θn is the phase of the nth harmonic. For random input noise, θn is not important and can be neglected. The total excess phase φ(t) and the ISF Fourier coefficients can be determined. The excess phase caused by any arbitrary noise current can then be calculated. Reference [3] also shows that there is frequency translation of noise components due to LTV behavior of noise in the oscillator. Consider a sinusoidal current at a frequency offset Δω from $m\omega_0$ injected into the system. The current can be expressed as

$$i(t) = I_m \cos((m\omega_o + \Delta\omega)t) \tag{18}$$

Neglecting the contribution from integrals with n ≠ m, the excess phase now reduces to

$$\phi(t) = \frac{I_m c_m \sin(\Delta\omega t)}{2q_{\max}\Delta\omega} \tag{19}$$

Thus the spectrum of φ(t) has sidebands at ± Δω even though the injected noise current was at mω0. Finally, for a sinusoidal oscillator with output $V_0(t) = V_0 \cos[\omega_0 t + \varphi(t)]$, the noise sidebands are given by

$$P_{SBC}(\Delta\omega) \approx 10\log_{10}\left(\frac{I_m c_m}{4q_{\max}\Delta\omega}\right)^2 \tag{20}$$

For a white noise source, $I_n^2/2$ is replaced by $i_n^2/(\Delta f)$, and the SSB power spectral density of the phase noise is

$$L(\Delta\omega) = 10\log_{10}\left\{\frac{\frac{i_n^2}{\Delta f}\sum_{n=1}^{\infty} c_n^2}{4q_{\max}^2\Delta\omega^2}\right\} \tag{21}$$

Parseval's relation can be used to express this quantity in terms of Γ_{rms}, which is the root mean squared value of Γ (x):

$$L(\Delta\omega) = 10\log_{10}\left\{\frac{\frac{i_n^2}{\Delta f}}{2\Delta\omega^2}\cdot\frac{\Gamma_{rms}^2}{q_{\max}^2}\right\} \tag{22}$$

At very low frequency offsets, in the flicker noise dominated portion of the noise spectrum ($\Delta\omega < \omega_{1/f}$), the device noise can be expressed as

$$i_{n,1/f} = i_n\left(\frac{\omega_{1/f}}{\Delta\omega}\right) \tag{23}$$

where $\omega_{1/f}$ is the device 1/f noise corner. This leads to the following expression for SSB phase noise in the $1/f^3$ region:

$$L(\Delta\omega) = 10\log_{10}\left\{\frac{\frac{i_n^2}{\Delta f}}{8\Delta\omega^2}\cdot\frac{c_o^2}{q_{\max}^2}\cdot\frac{\omega_{1/f}}{\Delta\omega}\right\} \tag{24}$$

where $c_o{}^2$ is the DC term in the Fourier series expansion of Γ (x). Thus, if c_0, the DC value of the ISF is a very small value, up-conversion of flicker noise is minimized. Therefore, we should design for an output waveform with small c_0, which is achieved when the output waveform has symmetric rise and fall times. Finally, the $1/f^3$ corner frequency is related to the 1/f corner frequency through

$$\omega_{1/f^3} = \omega_{1/f}\cdot\left(\frac{c_o^2}{4\Gamma_{rms}^2}\right) \tag{25}$$

To summarize, an accurate quantitative expression for the phase noise of a VCO has been developed in [3]. This theory also relates upconversion of 1/f noise to the symmetry of the waveform and predicts the precise expression for the $1/f^3$ noise corner. This new phase noise theory explains all mechanisms by which noise converts to phase noise in all regions of the phase noise plot.

3 Millimeter Wave VCO Design

In the early days of millimeter wave IC design, resonators were synthesized using distributed components such as compact slow-wave transmission line structures. However, as the field continues to mature, lumped element realizations have become more common. Millimeter wave VCOs require small tank inductors due to large parasitic capacitances contributed by MOS switches and cross-coupled pair. This significantly increases the effect of routing inductance on uniformity of tuning range [8]. If the entire tuning range is obtained by a single accumulation-mode (AMOS) varactor, the phase noise performance of wide-band LC-VCOs is significantly degraded due to AM-PM conversion [9]. Coarse tuning is performed in RF and mm wave VCOs using a switched capacitor bank (SWCB) having a binary implementation and a differential structure for better quality factor (Q) [10]. Thermometer implementation of the SWCB is very difficult in mm wave VCOs due to increased parasitic capacitances. Binary weighted SWCBs potentially have nonuniform and/or nonmonotonic steps. Nonuniform tuning range causes large frequency steps for certain capacitance codes leading to a large varactor size to cover the frequency spacing between coarse tune levels. Design of mm wave VCOs is quite challenging due to large parasitic capacitances contributed by the spiral inductor, MOS-FET cross-coupled pair and switched capacitor bank devices. There are further challenges associated with obtaining uniform tuning ranges at mm wave band frequencies due to significant routing parasitics. Reference [8] analyzes the challenges associated with discrete tuning range design in mm wave oscillators and proposes an architecture to obtain uniform tuning range. Small nonmonotonicities can be easily handled by a digital coarse-tuning loop. However, any large nonmonotonicity will cause an overall tuning range penalty. To avoid large nonmonotonic steps, the SWCB should be implemented in a symmetric fashion. Large frequency steps are caused mainly due to higher effective parasitic inductance associated with certain capacitance codes. To keep the frequency steps uniform across capacitance code, effective inductance associated with each capacitance code must be same. Each capacitor along with its self-routing inductance is broken into two halves and laid out symmetrically. The frequency of oscillation for such a distribution is given by Eq. (26) and Eq. (27) below.

$$F_{Q,OSC} \approx \frac{1}{\sqrt{L_T(C_T + C_{SW,Q})}}$$

where

$$C_{SW,Q} = C_{U,OFF}.\{F_Q(N-1) + F_Q(N-2) + \cdots + F_Q(0)\} \tag{26}$$

where

$$F_Q(i) = \left(1 + \frac{L_{PS,i} + L_{PR,i}}{2.L_T}\right) \cdot F(i)$$

$$\begin{aligned} L_{PS,i} + L_{PR,i} &= \sum_{n=i}^{N-1} L_{PS,n} + \sum_{n=0}^{N-1} L_{PS,n} + \sum_{n=0}^{i} L_{PS,n} \\ &= 2\sum_{n=0}^{N-1} L_{PS,n} \approx 2 \cdot \sum_{n=0}^{N-1} 2^{n-1} L_{PS,0} \end{aligned} \tag{27}$$

Here, L_T denotes the tank inductance, C_T the fixed tank capacitance, and CSW the total SWCB capacitance of the SWCB. F (i) represents the contribution from the ith bit capacitor of the overall SWCB in ON and OFF conditions. $C_{U,OFF}$ denotes the capacitance from the LSB in OFF condition. At millimeter wave frequencies, modeling of routing inductance becomes important, and the tuning range profile across capacitance code varies according to the relative position of the individual capacitors with respect to the tank inductance (L_T). The differential current flowing through each switched capacitor unit passes through tank inductance L_T, its self-routing inductance $L_{PS,i}$ and extra routing inductance $L_{PR,i}$. Self-routing inductance ($L_{PS,i}$) is the equivalent inductance that is added by the ith switchable capacitor. $L_{PS,0}$ is the self-routing inductance associated with the L_{SB}. Extra routing inductance ($L_{PR,i}$) is the equivalent inductance that is added between the *i*th switchable capacitor and tank inductance (LT) due to placement of other switchable capacitors. Each capacitor sees a total inductance equal to the sum of L_T, $L_{PS,i}$ and $L_{PR,i}$. Since larger capacitor banks occupy more area, they tend to have larger parasitic inductances. The symmetric SWCB topology proposed in [8] provides uniform tuning range even in the presence of significant routing inductance, which makes it an appropriate choice for the tank in a mm wave oscillator. This is validated in [8] through the fabrication and measurement of two oscillators with 5-bit switched capacitor banks, in a 130 nm SiGe BiCMOS process. In fact, it is shown that complete symmetrical distribution of SWCB is not always necessary, as EM simulations indicate that the self-routing inductance associated with the lower bits has minimal impact on their effective capacitance. Each VCO occupies a total area of 0.48 mm^2 and utilizes a differential center-tap 50pH inductor. The overall tank quality factor is about 14.5. A comparatively smaller tank inductance is chosen to maximize the tuning range in the presence of large parasitic capacitance of the BJT cross-coupled pair. The maximum frequency variation between adjacent capacitance codes for the control oscillator is 3.6× higher than that for the oscillator with the proposed SWCB layout. This leads to a 3.6× reduction in required varactor size, which ultimately results in improved VCO phase noise.

4 Millimeter Wave Class-C VCOs

The noise from the LC-tank and cross-coupled pair in a VCO cannot be reduced beyond a certain point. The tuning range specification places a strong constraint on high-Q oscillator design in a given technology. Under these conditions, the only way to improve phase noise performance is through reduction of bias current noise and increase output voltage swing. It has been shown previously in [11] that a class-C VCO gives higher voltage amplitudes and hence better phase noise performance for a given bias current. To further enhance performance, several class-C VCO architectures in CMOS technologies have been proposed in the literature. A 19.2 GHz bipolar class-C VCO is implemented in a 130 nm BiCMOS process in [12], with associated phase noise analysis. Specifically, it introduces analysis of Class-C VCOs with additional delay in the oscillator loop. This provides a dynamic biasing technique to achieve maximum oscillation swing without the requirement for a high unity gain bandwidth op-amp and also eliminates bias current noise.

A degeneration resistor (R_E) is used with every BJT to enable better current control in the presence of mismatches. The resistively degenerated bipolar transistor is henceforth referred to as a BJT unit. The I–V characteristics of the BJT unit can be modeled with good accuracy using a piecewise linear approximation, where $V\gamma$ is the cut-in voltage and $g_m = \frac{dI_c}{dV_{BP}}|_{DC}$ is the effective DC transconductance. $V_{B,DC}$ and $V_{B,SS}$ are the average voltages at the base terminal of the BJT unit in the absence and presence of oscillation, respectively.

During the start-up period, oscillation amplitude is small enough to operate in the linear portion of the characteristic around $V_{B,DC}$. Gradually, as oscillation amplitude grows, the BJT units turn off for certain portions of the oscillation period, when the instantaneous voltage goes below the cut-in voltage. As a result, the current injected by the cross-coupled pair into the tank over one period is higher than its DC current. The dynamic bias feature of the circuit reduces average value of voltage V_{BP}, leading to lesser injection of current into the tank. When the current injected into the tank over one cycle becomes equal to its DC current, the circuit reaches a periodic steady state. This forces the average base voltage ($V_{B,SS}$) to be lower than its DC value ($V_{B,DC}$), and each BJT unit operates with a certain conduction angle θ.

4.1 Amplitude of Oscillation

Assume that the circuit conditions are such that the oscillator has achieved periodic steady state with $V_{OP} = A_{OSC} \cos\phi$. The total current injected into the tank is

$$I_{TP} = N \cdot g_m \cdot [(V_{B,SS} - V_\gamma) + \alpha_o \cdot A_{OSC} \cdot \cos\phi] \quad \text{For } \phi, \text{ where } I_{TP} > 0 \quad = 0, \text{ otherwise} \tag{28}$$

where N is the number of BJT units, $\alpha_o = \frac{C_F}{C_F + C_B}$ is the feedback factor from V_{ON} to V_{BP}, A_{OSC} is the amplitude of oscillation, oscillation frequency $f_o = \frac{1}{2\pi\sqrt{L_T(C_T + \frac{C_F.C_B}{C_F.+C_B})}}$ and $\phi = 2\pi f_o t$ I_{TP} should be equal to 0 when φ is equal to the conduction angle θ, which leads to

$$\cos\theta = \frac{-(V_{B,SS} - V_\gamma)}{\alpha_o.A_{OSC}} \tag{29}$$

The expression for I_{TP} can then be rewritten in the following way

$$I_{TP} = N \cdot g_m \cdot \alpha_o \cdot A_{OSC} \cdot [\cos\phi - \cos\theta] \tag{30}$$

for $-\theta \le \varphi \le \theta$
= 0, otherwise

The average value of I_{TP} over one oscillation period in periodic steady state must be the same as the DC current flowing through the tank, which is equal to ($\frac{N.I_{Bias}}{2}$). Therefore, the expression for the tank bias current is

$$I_{Bias} = 2 \cdot \frac{g_m \cdot \alpha_o \cdot A_{OSC} \cdot [\sin\theta - \theta \cdot \cos\theta]}{\pi} \tag{31}$$

The fundamental component of I_{TP} can be written as

$$I_{TP,1} = \frac{1}{\pi} N \cdot g_m \cdot \alpha_o \cdot A_{OSC} \cdot [\theta - 0.5 \cdot \sin 2\theta] \tag{32}$$

Based on Eq. (31) and Eq. (32), the fundamental component of I_{TP} is given by

$$I_{TP,1} = \frac{N.I_{Bias}}{2} \cdot \left[\frac{\theta - 0.5.\sin 2\theta}{\sin\theta - \theta.\cos\theta}\right] \tag{33}$$

Assuming that the tank quality factor (Q) is large enough to filter out higher harmonics of I_{TP}, the amplitude of oscillation A_{OSC} is

$$A_{OSC} = \frac{N \cdot I_{Bias} \cdot R_T}{2} \cdot \left[\frac{\theta - 0.5 \cdot \sin 2\theta}{\sin\theta - \theta \cdot \cos\theta}\right] \tag{34}$$

where R_T represents the equivalent parallel resistance of the tank. Equation (34) shows that amplitude of oscillation increases with decrease in conduction angle. The theoretical maximum amplitude of oscillation is obtained when the conduction angle is 0°, which is not achievable in a practical circuit. It is useful to understand the limitation on minimum possible conduction angle. Consequently, θ is related to the circuit parameters through the VCO start-up gain (V SG_o) in the following way

$$V_{SG_o} = N \cdot g_m \cdot \alpha_o \cdot R_T = \frac{\pi}{\theta - 0.5 \cdot \sin 2\theta} \tag{35}$$

when plotted, Eq. (35) shows that requirement of VCO start-up gain increases very rapidly for small conduction angles. For example, a conduction angle of 45° already requires a startup gain of 11, which is very difficult to achieve in a practical oscillator. Given the typical limitations on tank quality factor and supply voltage, the VCO start-up gain is usually limited to around 5, which corresponds to a conduction angle of approximately 60°. It should also be noted that Class-C oscillators potentially have amplitude instability of oscillation, represented by low frequency variation in oscillation amplitude. The tail resistance $R_{2,TOT}$ provides the DC current path and it has to be higher than a certain value to ensure amplitude stability of oscillation [13].

4.2 Phase Noise

For a differential LC oscillator, the phase noise L(Δω) at an offset frequency Δω from the center frequency is given by

$$L(\Delta\omega) = 10\mathrm{L}og\left\{2 \cdot \frac{N_{TK} + N_{CC}}{A_{OSC}^2/2}\right\} \tag{36}$$

where A_{OSC} is the amplitude of oscillation from Eq. (34). N_{TK} and N_{CC} are the effective noise due to the tank and cross-coupled pair respectively, and their expressions are as follows:

$$N_{TK} = \frac{\Gamma_{TK}^2}{4\Delta\omega^2 C_{TOT}^2} \cdot \overline{i_{n,TK}^2}/\Delta f \tag{37}$$

$$N_{CC} = \frac{\Gamma_{CC}^2}{4\Delta\omega^2 C_{TOT}^2} \cdot \overline{i_{n,CC}^2}/\Delta f \tag{38}$$

here, Γ_{TK} and Γ_{CC} are the ISF (Impulse Sensitivity Function) for the tank noise current and cross-coupled pair noise current, respectively; $\overline{i_{n,TK}^2}/\Delta f$ and $\overline{i_{n,CC}^2}/\Delta f$ are the tank current noise and cross-coupled pair current noise spectral densities respectively; and, C_{TOT} is the total tank capacitance given by $(C_T + \frac{C_F.C_B}{C_F+C_B})$ We can now write the expressions for the noise currents in the following way

$$\overline{i_{n,TK}^2} = \frac{4k_B T}{R_T} \tag{39}$$

$$\overline{i_{n,CC}^2} = 4k_B T \cdot \gamma \cdot g_m(\phi) \tag{40}$$

where R_T is the equivalent parallel tank resistor; gm(φ) is the transconductance of the cross-coupled pair; γ is the effective excess noise factor of the cross-coupled devices.

In the above expression for phase noise in Eq. (36), we assume that thermal noise of the tank and cross-coupled pair are the only two noise sources in the circuit. First, bias current noise is removed using the large off-chip capacitor C_{EXT}, while thermal noise of $R_{2,TOT}$ at twice the oscillation frequency is eliminated through filtering by shunt capacitance $C_{2,TOT}$. Finally, to a large extent, it can be shown that the flicker noise of the cross-coupled pair does not contribute to phase noise. A closed-form expression for the phase noise of the bipolar class-C oscillator may be derived as in Eq. (41) below, using piecewise linear characteristics of the BJT unit cell-based cross-coupled pair. This also confirms that phase noise does not depend on cross-coupled pair devices or current pulse shape, as expected for this class-C oscillator.

$$L(\Delta\omega) = 10 \cdot Log\left(\frac{k_B T\left(1 + \frac{\gamma}{\alpha_o}\right)}{2.\Delta\omega^2 \cdot C_{TOT}^2 \cdot A_{OSC}^2 \cdot R_T}\right) \tag{41}$$

4.3 *Ideal Phase Delay in Oscillator Loop*

During actual implementation of high-frequency oscillators, it is expected that there will be finite delays in the oscillator loop due to device and routing parasitics. These will be all the more important in mm wave designs, where delays will be a non-insignificant portion of the period of the oscillation waveform. To more easily analyze how this affects the performance of a mm wave oscillator, we can lump it together into a single delay element in the feedback path. Ideal phase delay (φ_o) represents the effective time delay at the frequency of oscillation (a unity gain buffer is added to simplify loop gain analysis). Loading on the tank due to C_F and C_B has been modeled by including them as a part of the LC tank. A phase delay of ($\angle - \varphi_o$) in the oscillator loop forces it to oscillate at a different frequency (f_D) compared with the original oscillation frequency (f_0), such that it provides an exact but opposite phase delay ($\angle + \varphi_o$) to sustain the oscillation. The new oscillation frequency f_D is determined by the fact that, at the oscillation frequency, overall phase $\angle$ $LG_D(j\omega)$ in the loop must be zero. Here, we can write the small signal loop gain as

$$LG_g(j\omega) = N \cdot g_m \cdot \frac{1}{\{Y_T(j\omega) + Y_{FD}(j\omega)\}} \cdot \alpha_o \angle e^{-j\phi_o} \tag{42a}$$

where, $Y_T(j\omega) = \frac{1}{j\omega.L_T} + \frac{1}{R_T} + j\omega \cdot C_T$

$$Y_{FD}(j\omega) = j\omega \cdot (\frac{C_F \cdot C_B}{C_F + C_B}) \tag{42b}$$

The magnitude of tank impedance at f_D, denoted by $|ZD|$, is

$$|Z_D| = |\frac{1}{\{Y_T(j\omega) + Y_{FD}(j\omega)\}}|_{\omega=2\pi f_D} \tag{43}$$

Once the circuit achieves periodic steady state in the presence of ideal phase delay (φ_o) in the oscillation loop, we can write

$$V_{OP}(\phi) = -A_{OSC,D} . \cos(\phi + \phi_o)$$

$$V_{BP}(\phi) = \alpha_o \cdot A_{OSC,D} \cdot \cos(\phi) \tag{44}$$

where, $A_{OSC,D}$ is the oscillation amplitude and $\varphi = 2\pi\ f_D\ t$. Following the analysis presented earlier for the case without any delay, the oscillation amplitude in the presence of ideal phase delay can be derived to be

$$A_{OSC,D} = \frac{N \cdot I_{Bias}}{2} \cdot |Z_D| \cdot \frac{(\theta_D - 0.5 \cdot \sin 2\theta_D)}{(\sin\theta_D - \theta_D \cdot \cos\theta_D)} \tag{45}$$

Similarly, the expression for start-up gain ($V\ SG_D$) in terms of conduction angle (θ_D) in the presence of ideal phase delay is

$$VSG_D = N \cdot g_m \cdot \alpha_o |Z_D| = \frac{\pi}{(\theta_D - 0.5 \cdot \sin 2\theta_D)} \tag{46}$$

Since the oscillation takes place away from the resonant peak of the stand-alone LC tank, $|Z_D|$ is smaller than the original equivalent tank parallel resistance (R_T). This reduces start-up gain, increases conduction angle and ultimately degrades the oscillation amplitude for the same amount of bias current injected into the tank. Following the analysis presented earlier for the case without any delay, phase noise of an LC oscillator in the presence of ideal phase delay in the oscillator loop is given by

$$L_D(\Delta\omega) = 10\mathrm{L}og \left\{ 2 \cdot \frac{N_{TK,D} + N_{CC,D}}{A^2_{OSC,D}/2} \right\} \tag{47}$$

where the respective quantities are indicated by appropriate subscripts. Since the oscillation is offset from the LC tank resonance, we expect the noise contribution of the tank to be higher in the presence of delay due to reduced amplitude for the same bias current. In other words, tank current noise remains the same, but its contribution to phase noise increases compared to the original case. In fact, the tank ISF in the presence of ideal loop delay, $\Gamma_{TK,D}(\phi)$, and its RMS value $\overline{\Gamma^2_{TK,D}(\phi)}$ can be derived as follows,

$$\Gamma_{TK,D}(\phi) \approx \frac{R_T}{|Z_D|} \cdot \frac{\sin(\phi)}{2} \tag{48}$$

$$\overline{\Gamma^2_{TK,D}(\phi)} = \frac{1}{8} \cdot \frac{R_T^2}{|Z_D|^2} \tag{49}$$

Noise contribution of the tank ($N_{TK,D}$) in the presence of ideal phase delay can be derived using Eq. (37), Eq. (39) and Eq. (49) as follows

$$N_{TK,D} = \frac{1}{8.\Delta\omega^2 \cdot C_{TOT}^2} \cdot \frac{k_B T}{R_T} \cdot \frac{R_T^2}{|Z_D|^2} \tag{50}$$

Cross-coupled pair ISF ($\Gamma'_{CC,D}(\varphi)$) and its RMS value $\overline{\Gamma'^2_{CC,D}}$ in the presence of ideal loop delay are

$$\Gamma'^2_{CC,D}(\phi) = \frac{R_T}{|Z_D|} \cdot \frac{\sin(\phi)}{2} \cdot \beta_D(\phi) \tag{51}$$

where,

$\beta_D^2(\phi) = 1$ for $-\theta_D < \varphi < \theta_D$,
$= 0$, otherwise (52)

$$\overline{\Gamma'^2_{CC,D}(\phi)} = \frac{1}{8\pi} \cdot \frac{R_T^2}{|Z_D|^2} \cdot (\theta_D - 0.5 \cdot \sin 2\theta_D) \tag{53}$$

Finally, noise contribution from the cross-coupled pair ($N_{CC,D}$) in the presence of ideal loop delay can be derived using from Eq. (46) and Eq. (53) as follows (54)

$$N_{CC,D} = \frac{1}{8.\Delta\omega^2 \cdot C_{TOT}^2} \cdot \frac{\gamma}{\alpha_o} \cdot \frac{k_B T}{|Z_D|} \cdot \frac{R_T^2}{|Z_D|^2} \tag{54}$$

In summary, phase noise of a differential LC oscillator in the presence of ideal loop delay is given by

$$L_D(\Delta\omega) = 10\mathrm{Log}\left\{\frac{k_B T \cdot (1 + \frac{\gamma}{\alpha_o} \cdot \frac{R_T}{|Z_D|})\frac{R_T^2}{|Z_D|^2}}{2.\Delta\omega^2 \cdot C_{TOT}^2 \cdot A_{OSD,D}^2 \cdot R_T}\right\} \tag{55}$$

5 Bipolar Class-C VCO with Device Parasitics

We now very briefly consider how to handle analysis including internal parasitics of the BJT unit used in the cross-coupled pair, with output resistance r_o, collector-emitter capacitance c_{ce}, base-collector capacitance c_{bc}, base-emitter capacitance c_{be} and base resistance r_b. After optimization of device parasitics, overall phase delay introduced in the loop can be considered for phase noise analysis as follows. First, we note that the emitter terminal of the BJT experiences a much smaller signal swing compared with the collector terminal, when operating in class-C mode. Hence, the effect of r_o and c_{ce} can be absorbed into that of the LC tank. Similarly, the effect of c_{bc} can be considered in the form of a reduced effective feedback factor (αA). Moreover, c_{be} and r_b are the major parasitics that degrade the performance of the oscillator in most bipolar processes. To understand their effect quantitatively, we can consider a simplified model of the BJT unit cell, where c_{be} is connected between base and signal ground, considering the low swing at the emitter node. It can be shown that R_B not only introduces delay in the loop, but also reduces the magnitude of αA. Noise of R_B appears as part of both the tank and feedback networks due to loading effects, but with contribution to the feedback network being significantly filtered by C_F and C_B. Therefore, this can be effectively ignored, and noise of RB is considered only for tank noise calculations. Let $R_{T,EQ}$ be the effective tank parallel resistance at the natural resonant frequency of the tank. In summary, we can write the expression for oscillator phase noise in the presence of device and routing parasitics as follows.

$$L_A(\Delta\omega) = 10Log\left\{\frac{k_B T \cdot \left(1 + \frac{\gamma}{\alpha_o} \cdot \frac{R_{T,EQ}}{|Z_A|}\right)\frac{R_{T,EQ}^2}{|Z_A|^2}}{2.\Delta\omega^2 \cdot C_{TOT}^2 \cdot A_{OSD,A}^2 \cdot R_{T,EQ}}\right\} \tag{56}$$

where the subscript A for each variable refers to the case when they are derived for the device with actual parasitics, as per the analysis presented earlier in this chapter. Ref. [12] then compares the cases with ideal delay and actual bipolar parasitics, and demonstrates that the phase noise degradation due to delay caused by device and routing parasitics is much worse than the same amount of ideal delay in the loop. This is expected as the presence of base resistance loads the tank and reduces overall quality factor.

The analysis presented so far models delay in the oscillator loop with lumped components. In reality, delay can also be caused due to distributed effects. Replacement of a lumped network with a distributed network reduces the overall phase delay in the oscillator loop and improves quality factor of the tank. The case with a distributed feedback network is expected to exhibit lesser phase noise degradation compared to one with a lumped feedback network. Ref. [12] validates the above analysis through comparison with circuit simulations and presents measured results on a 19.2 GHz VCO designed in a 130 nm BiCMOS process with the circuit topology proposed in [12].

6 Conclusion

This chapter introduced millimeter wave LC VCOs for 5G applications. The fundamentals of LC VCOs were described in detail, including an introduction to the basic operation of negative resistance oscillators and phase noise. De- sign considerations for LC-VCOs at 5G mm-wave frequencies were illustrated through two recently reported works. A proposed technique for monotonic discrete tuning of mm wave VCOs was first discussed. Then, circuit analysis and design techniques as applied to a K-band VCO were discussed in detail with a strongly analytical approach.

References

1. Razavi B (1998) RF microelectronics. Prentice Hall, New Jersey
2. Hajimiri A, Lee TH (1999) Design issues in CMOS differential LC oscillators. IEEE J Solid State Circuits (JSSC) 34(5):717–724
3. Hajimiri A, Lee TH (1998) A general theory of phase noise in electrical oscillators. IEEE J Solid State Circuits (JSSC) 33(2):179–194
4. Hajimiri A, Lee TH (2000) Oscillator phase noise: a tutorial. IEEE J Solid State Circuits (JSSC) 35(3):326–336
5. Leeson DB (1966) A simple model of feedback oscillator noise spectrum. Proc IEEE 54(2):329–330
6. Darabi H, Abidi AA (2000) Noise in RF-CMOS mixers: a simple physical model. IEEE J Solid State Circuits (JSSC) 35(1):15–25
7. Rael JJ, Abidi AA (2000) Physical processes of phase noise in differential LC oscillators. In: Proceedings of the custom integrated circuits conference, pp 569–572
8. Thakkar A, Theertham S, Mirajkar P, Aniruddhan S (2019) Techniques for improved continuous & discrete tuning range in millimeter wave VCOs. IEEE Trans Very Large Scale Integr (VLSI) Syst 27(3):729–733
9. Hegazi E, Abidi AA (2003) Varactor characteristics, oscillator tuning curves, and AM-FM conversion. IEEE J Solid State Circuits (JSSC) 38(6):1033–1039
10. Sjoland H (2002) Improved switched tuning of differential CMOS VCOs. IEEE Trans Circuits Syst II 49(5):352–355
11. Mazzanti A, Andreani P (2008) Class-C Harmonic CMOS VCOs, with a general result on phase noise. IEEE J Solid State Circuits (JSSC) 43(12):2716–2729

12. Thakkar A, Theertham S, Aniruddhan S (2018) Phase noise analysis of bipolar class-C VCOs with delay in oscillator loop. IEEE Trans Very Large Scale Integr (VLSI) Syst 26(12):2873–2883
13. Fanori L, Andreani P (2013) Highly efficient class-C CMOS VCOs, including a comparison with class-B VCOs. IEEE J Solid State Circuits (JSSC) 48(7):1730–1740

Mm-wave CMOS Power Amplifiers for 5G

Pradeep Gorre, Rajesh Kumar, Hanjung Song, and Sandeep Kumar

Abstract The chapter discusses the basic elements in the design of mm-wave CMOS Power Amplifier (PA) for phased arrays integration, focusing the next-generation 5G mobile communication. Power Amplifier design metrics, along with implementation of beam-forming phased arrays to merge power over-the-air are discussed in brief. The explanation begins with CMOS unique advantages, real-time handset challenges, system-level constraints, and design challenges are conceptually demonstrated with the help of a basic single-stage transistor Power Amplifier.

1 Silicon-Based PAs

Amplifier circuits are the key component in most of the electronic circuit designs that are intended to provide high power to simulate devices connected to its output. A Power Amplifier converts low input power supply to high output power. During the past years, RF power amplifiers (PAs) have gained tremendous attention especially in the development of wireless communication systems. But recently, broadband RF PAs are of great interest as there is strong consumer desire to enhance the functionality of wireless devices. This availability of broadband PAs would reduce the need for multiple amplifiers thus saving the development costs of chip area and speeding up the implementation. On the other hand, a reconfigurable feature has been demonstrated to be one of the most promising concepts in developing PA devices with high performance, low cost, small size and wide tuneable. In addition, it is seen that in matching networks, some transmission lines like quarter wave transformer, coplanar

P. Gorre · S. Kumar (✉)
Department of Electronics and Communication Engineering, National Institute of Technology Surathkal, Surathkal, Karnataka, India
e-mail: sandeep@nitk.edu.in

R. Kumar
Indian Institute of Technology Dhanbad, Dhanbad, Jharkhand, India

H. Song
Department of Nano Science and Engineering, Center for Nano Manufacturing, Inje University, Gimhae 621-749, Korea

S. Singh et al. (eds.), *CMOS Analog IC Design for 5G and Beyond*, Lecture Notes in Electrical Engineering 719, https://doi.org/10.1007/978-981-15-9865-4_7

waveguide, microstrip line, etc. have improved the broadband performance of power amplifiers instead of bulky inductor. Out of these, microstrip lines have enormous potential in achieving the circuit reconfigurability due to their superior performance and the fabrication compatibility with existing IC technology. The key performance metrics of Power Amplifier are linearity, power gain (Gp), output power P_{out}, and power-added efficiency (PAE). The saturated continuous-wave (CW) is measured as the product of maximum voltage and maximum current I_{max}. If P_{out} value is greater than P_{sat}, then the peaks of the signal at the Power Amplifier input are distorted at the output. Amplification gain Gp is the ratio of output power to input power. The transistor maximum available gain (MAG) is calculated at an optimized loading/matching criteria that boosts its overall gain.

2 Modes of PAs

2.1 Class A Operation

In this mode of operation, the transistor conducts in the linear region throughout the cycle. It signifies that the collector current operates for 360° (full cycle) which is given in [1]. For class A power amplifier, *Q* point generally locates at the midway of the load line. Hence, the output signal swings across the maximum possible range, which results a distorted waveform. The maximum efficiency (η) of a Class A power amplifier is 50%.

2.2 Class B Operation

In contrast to class A operation, here the output current operates only for one-half of the full cycle whose conduction angle is 180°, plotted in [1]. To achieve this, a design engineer maps the operating point in cut-off region. As, the transistor is switch off for half of the cycle, the transistor dissipates less power consumption, thus the η is enhanced. Class B power amplifier maximum efficiency is 78.5%.

2.3 Class C Operation

Another classification of power amplifier is class C operation, where the device conducts for less than half of the cycle and conduction angle of class C PA is less than 180° and its typical value lies between 80° and 120 which is referred in [1].

3 Advantages of Silicon-Based PAs

Silicon-based power amplifiers, in particular CMOS PAs, have currently originated as a best alternative solution in the design of power amplifiers with cheap, low static power and high-density production. They also feature like high noise immunity, fully integrated on-chip, achieving in a low footprint with a magnificent front-end device. In addition, CMOS PAs significantly integrate on a single system-on-chip (SoC), which makes it to meet the current industry trends in next-generation wireless communications. Especially, CMOS technology offers unsurpassed on-chip signal computation that eventually improves the RF PA performance, at the same time offering a best economical and minimum overhead. The advanced analog/digital signal processing methodologies may achieve best PA performance improvement, as in traditional digital pre-distortion (DPD) techniques taking advantage of all the particular benefits of CMOS technology to maximize the performance of PA which may ultimately result in a dramatic change in the PA design approaches. Consequently, sophisticated design approaches of CMOS PA have broaden their outlook from the basic RF system to a complex RF system with arranged combinations of analog and digital.

In preceding years, a continuous miniaturization of complementary metal oxide semiconductor (CMOS) technologies in nanometre scale is facing many reliability and degradation issues. Designer needs to pay more attention toward circuit design that is reliable and insensitive to the transistor parameter degradation. The resilient biasing technique aims to design reliable circuits that are capable of postprocess adjustment and insensitive to the transistor parameter degradation over long-term stress effect [1, 2]. Random doping fluctuation is one of the most important fluctuation sources for threshold voltage. Due to random doping profile, the fluctuation of threshold voltage is expected to be larger [3]. Threshold voltage variation is inherent to the property of CMOS. The sensitivity of the threshold voltage variations in the critical dimensions is greater due to increasing shorter channel effects as the gate length is reduced using CMOS technology.

4 Over-the-Air Power Combining: Phased Array Transmitters

Phased arrays technique effectively mitigates the losses in the transmission path despite minimum output power and transmitter directivity. In dual-polarized antennas, each antenna is driven by two transmitters, which increases signal-to-noise ratio in weak signal situations or allows polarization diversity at higher SNR and link budget of PA requirements are obtained from Friis formula. This equation justifies signal detection at the receiver end, which needs effective isotropic radiated power (EIRP) at transmitter to overcome integrated thermal noise received at the antenna, and also the noise contributions, with a low SNR govern by implementing effective modulation, i.e., SNR_{min}. Transmitter effective isotropic radiated power is defined as the total power radiated plus gain of the transmitter antenna. The important loss results in path loss and noise figure of the receiver. Further, transceiver thermal losses and implementation losses are considered in further section.

5 System considerations: Practical Handset Constraints

The phased array link budget of 5G broadband cellular communication system is evaluated to study a set of various potential carrier frequencies, along with user equipment (UE) transmitter power consumption *P*Tx as a parameter. The frequency at 28 GHz band that is an untapped frequency spectrum found to be a favorable choice for 5G cellular applications. The considerations for PA output power for upcoming 5G phased array PA are as follows.

5.1 *Choice of Carrier Frequency for 5G Systems*

In [2], the analysis plot for the link budget and loss mechanisms of the input signal in the direction of access point. By considering a line-of-sight channel clarify this study without comparing carrier frequencies. The choice of economical CMOS technology is preferred for the design of phased array RFIC, and a printed circuit board design antenna arrays is chosen. The arrangement of uniform rectangular patch antenna at every carrier frequency into an *Nx X Ny* dimensions to match UE/AP PCB dimensions *dx* ×*dy*, fixed across carrier frequency values at a spacing of 0.5 λ_0, where λ_0 is free space wavelength. The UE/AP URA at arbitrary f_c is preferred in [3]. Element counts versus f_c and an example UE array at $f_c = 30$ GHz are displayed in [3].

$$PTx, rf(f_c) = 10log_{10}\left(k_B Tb \times BW_{sig}\right) + NF_{Rx}(f_c) + SNR_{sig} + L_{path}(f_c) + L_{misc} + \left[L_{FE,Tx(f_c)} + L_{FE,Rx(f_c)}\right] - \left[G_{array,Tx(f_c)} + G_{array,Rx(f_c)}\right] \quad (1)$$

where k_B represents Boltzmann constant, *Tb* is absolute temperature, and P_{Tx}, rf (f_c) is the total power of the transmitter phased array in dBm, desired for signal detection. The basic condition is to demonstrate the best-tested functionalities for system equipment's from various publications over frequency range 2.4–83 GHz. The recent study on CMOS back-off PAE information shows approximated power added efficiency at *P*out that achieves |EVM| = SNR_{sig} + 3 dB = 25 dB with the help of 64-QAM OFDM by PAE at P_{sat} of −9.6 dB, then compared the results with the state-of-art.

$$PAE(f_c) \approx \frac{35\%}{\left[1 + 0.16\sqrt{\left(f_c/10^9\right)}\right]^2} \tag{2}$$

By considering the equations (1) and (2), the scattering of transmitter data and its trend line is given as *P*Tx, dc*(fc)* = *P*Tx, rf*(fc)*/PAE*(fc)*. Based on the recent publications, the best performances of *P*Tx, DC over a wide carrier frequency range is available in [3]. For finite range, the best modulation scheme is chosen along with bandwidth, and fixed antenna size. In [3], indicates efficient power enables at 28 GHz band in range 5–6 GHz even with minimum PAE and maximum loss in propagation.

$$\frac{PTx, rf(f_c)}{PAE(f_c)} \propto \left[\frac{L_{path,A(f_c)}}{G_{array,Tx(f_c)}G_{array,Rx(f_c)}}\right].\frac{1}{PAE(f_c)} \sim \left[\frac{f_c^2}{f_c^2 \times f_c^2}\right].\frac{1}{\left(1/f_c\right)} \approx \frac{1}{f_c} \tag{3}$$

Equation (3) signifies the impact on using directive arrays at mm-Wave frequencies for both access point and end user to overcome the losses in the transmission path suffer at low carrier frequency, where the dimensions of antenna limit the number of antenna elements. The combined gains of phased array antennas at transmitter and receiver over the frequency range 2.4–15 GHz are shown in [3], which are inadequate to mitigate the adverse impact of rising path and decline behavior of PAE on *P*Tx, DC. In addition, the red cultured portion in figure available specifies that *L*FE and NF *Rx* in (1) might restrict the gain of antenna arrays. Hence, the discussed semiempirical study clears that 28 GHz frequency band is a best suitable frequency for broadband 5G applications (Table 1).

Table 1 Summary of targeted PA Circuit specifications

PA circuit specification	Value (unit)	Comment
Signal Format RF Bandwidth BWsig PAPR	64-QAM OFDM 250 (MHz) 9.6 dB	Most challenging 5G mm-wave uplink use case
Required Transmitter EVM, i.e., EVM_{req}	−25 (dBc)	3dB margin from SNRsig = 22 dB for 64-QAM detection
Required average Pout @ EVMreq, i.e., Pout,req	7.0 (dBm)	≈100 m estimate from relatively optimistic link budget based on e.g., [5]
Small-signal Gain, Gss	≥ 15.0 (**dB**)	To avoid degrading transmit chain DC power consumption [6]

5.2 Output Power Requirements

The efficient uplink scheme makes use of a 250 MHz ultra-wide 64-QAM OFDM signal. In the present research on mm-wave frequency especially at 28 GHz, the work focuses on real-time transmission range and *P*Tx, DC which is a challenging task. By taking size limitation and signal losses into consideration from the signal analysis is shown in [3]. The required average *P*out at 30 GHz/UE is illustrated in figure given as [3] which is intended to obtain a range of 10–150 m line-of-sight. To establish 20–30 m line-of-sight, it desires an average *P*out of 5–7.5 dBm at 250 MHz frequency. On the other hand, by considering the 16-element phased array establish a 50 m range with average *P*out nearly equal to 7 dBm. In case of low power 5G, it requires a longer range, which can employ lower modulation scenario like QPSK is shown in [3]. An important point to observe is the error vector magnitude is a key metric, which quantifies the linearity of the system and should meet the requirements in the design of mm-wave CMOS PA. Adjacent channel power ratio is given minimum priority as its spatial selectivity is high at mm-wave frequencies when compared with the microwave frequency range. It is also noticed that, at 6 GHz frequency, adjacent channel power ratio of one path blocks the adjacent channel signal received which tends antenna to fail in separation of channels. Alternatively, in the present situation phased arrays at mm-wave are enough to separate spatially. If two or more users utilizing the same coverage sector that are maintained under single beam from an access point, their interference could be alleviating by decreasing their bandwidths. In general, these scenarios occur occasionally.

6 Power Gain Requirements

Power gain requirement of a PA's is set by output power, chosen device technology, chip area availability, and maximum input power. The intermediate stage between power amplifier and converters in a transmitter insert many stages, which operate at low frequencies, so that PA input power is restricted mostly by the driver circuit, which could be a phase shifter employed in RF phased arrays, or a mixer circuit in local oscillator phase shifting arrays. The safety measurements of PA input power in the mobile edge schemes are up to -10 dBm while, maximum throughput up to $-$20 dBm. Hence, cascade topologies are preferred to design a bulk CMOS PA.

7 Linearity Requirements

The main requirement of an amplifier is to generate an output signal, which follows the input signal. The efficiency of millimeter-wave power amplifiers can be improved by considering Class B power amplifiers, which results a high efficiency when compared Class A amplifiers. However, Class B and Class AB operation result in large crossover distortion, which degrades the ACPR and EVM. The AM-PM distortion that occurs at various voltage biasing is ref from [3]. It is observed that a minimum distortion is achieved by considering an optimum bias point. Yet, this bias technique is not good choice while considering efficiency of the PA and output. A two-stage common-source PA topology employs various bias voltages to satisfy the complete AM-PM distortion. The inductor degeneration circuit results a better linearity. SNR_{min} specifies the overall achievable Tx EVM. The complete transmitter design components and their contributions to budget can be evaluated prior. The estimation of SNR budget at Mm-Wave frequencies plays a dominant role in the design of linear CMOS PAs. The budgeting is one of the parameters to be taken into consideration and the PA performance for Pout refers to one-half of the total budget. In design considerations, the convenient PA EVM role is chosen appropriate to this estimate, however, the designer calculates the specified target Tx requirements to achieve an adequate spectral efficiency.

8 Mm-Wave CMOS PA Challenges

8.1 Switching Versus Linear PAs

A simple power amplifier can be broadly classified as "switching" or "linear" power amplifiers. The transistor in switching PAs behaves like a voltage-control source, where output is measured at DC levels. RF-range switching PAs achieve more effective waveform shaping and high PAE value but, these power amplifiers are inconsistent to spectrally efficient. In addition, the bandwidth range for 5G NR is 50–800 MHz and present conditions overtake the state-of-the-art modulators employed in recent architectures like polar transmitter, where a QAM technique is used in switching Power Amplifiers. In addition, losses in passives at Mm-Wave frequencies block higher harmonics, shrinking the efficiency benefits of switching Power Amplifiers. The Linear Power Amplifiers are highly reliable in current conditions, as the active device of the power amplifier acts as a current–control source.

8.2 Single-Transistor Linear PA

The block diagram of the signal flow representation of a single flow CMOS PA circuit is shown in [2]. The intended forward path versus parasitic feedback path is also illustrated in [2]. A conventional and highly simplified design approach to attain an average output power is:

1. Choose a suitable process technology, i.e., short-channel technology achieves faster transitions but is more expensive.

2. Select a proper VDD for the selected channel length; trade-off between attainable Pout and reliability.
3. Select proper bias technique; to increase f_{max} in a reasonable process technology.
4. Select a proper channel width to sustain P_{sat} as a minimum Pout plus PAPR (avoid clipping). To attain high Pout, it requires a wider gate, however, this affects f_{max} by boosting parasitic feedback.
5. Finally, design an impedance matching circuit at output to match impedance of antenna Z_Ant with finest load impedance of PA. The CMOS PAs are integrated into multiple phased arrays for 5G NR, where the cost of fabrication drives Zopt calculating by means of load-pull simulations instead of experiments.

8.3 Single Stage Class AB PA with Tuneable Body Biasing

The design is a resilient biasing technique for PA using silicon substrate with 65 nm PTM and achieved a more stable design in terms of sensitivity of threshold voltage variation. During analysis, it is found that the level of reduction in threshold voltage of MOSFET is related to the body effect coefficient and MOSFET structure coefficient.

- ***Output Power***

Proper selection of VDD limits output power more than maximum current in CMOS, intensify by scaling. A transistor attains a low knee voltage at peak current, which drains far from VDD. The scaling of knee voltage is not possible, but dependability compels VDD to minimize along with gate length. Periodically, scaling increases gain but lessen VDD

In addition, the mobility of user equipment involves arbitrary proximity to conducting elements, hence antenna impedance (Z_Ant) changes randomly time. At high frequencies, antenna impedance mismatch is compared with impedance of feed-line. In favor of minimal mismatches, varying Z_Ant plot to an altering PA load over its simple value Zopt, so P_{out} arbitrarily decreases. It is a complex task to maintain signal integrity at Mm-wave frequency, so output matching will almost integrated on chip. High losses in power occur in CMOS network design due to its high substrate conductivity Additionally, obtaining high Psat results low VDD, which leads to a sharper output network transformation to map Z_Ant to a lower Zopt, which also add loss because of complexity in matching network.

The above challenges of high-power CMOS PA known, but these challenging are to handle with more attention at Mm-wave frequency for following reasons. First, restriction on number of antennas in the mobile devices percept large Pout per PA. Output power measured is less when compared with single-PA case, while considering the required fmax of the device, it selects the cutting-edge process. Second, the on-chip integration of transmitter-receiver antenna duplexer implements time-division duplexing that adds significant output losses. Third, implementation of phased array architecture in broadband communication is a new approach for 5G NR.

- **Power Gain**

The gain of the amplifier is limited by two issues. First, transistor-to-transistor and transistor-to-passive components wiring naturally introduce an extrinsic parasitic capacitances and internal resistances. Therefore, a "fully-wired" intrinsic value is remarkably higher than the value of f_{max}. In practice, f_{max} value drops to 150 GHz (250 GHz) in a fully wired transistor. This effect has to consider if power gain at PA output reaches less than 6–8 dB, while the desired driver stage is fundamentally a PA. Hence, the design flow has to begin with physical layout design of transistor and carefully adjust intrinsic and extrinsic capacitance parasitic and also f_{max} value depends on voltage bias, but an optimal selection of gate bias is more challenging than the process discusses above; which is a trade-off for efficiency, linearity, PAE and f_{max}.

- **Power-Added Efficiency**

The digital-to-analog conversion mechanism in a PA establishes a high PAE at P_{sat} only, whereas signal EVM and PAPR requirements induce considerable back-off. In [2] shows a literature survey on power-efficient PAs, where PAE -falls significantly at back-off values of 6, 8 or 10 dB. PAE is one of the design metric of PA analyzed in prior sections, which may be centralized into four factors:

- Device technology, gate bias and width, and waveform engineering.
- Degrades with lowering VDD and gate length.
- It is a process technology, gate-bias and layout dependent.
- Decreases with power loss at output, in particular for higher Pout targets that require steeper matching networks.

- **Linearity**

The behavior of the transistor nonlinearity characteristics and intrinsic values of gate capacitances C_{gs} and C_{gd} are shown in figure that ref from [2], which depend on spontaneous terminal voltages. The output curves of modulation are flat for a perfectly linear PA, and for real-time PA.

The capacitances, and interactions with passive components. Distortion can be limited by increasing back-off; hence the peaks of the modulated signal pass through amplitude and phase modulation flat curves. Yet, these high values of back-off adversely limit PAE as illustrated in [2]. In addition, transistor nonlinear characteristics of voltage and current, and transistor capacitances rely on cumulative of past and present data of its terminal values of voltages. Previous terminal values are due to the echoes of the signal, which stay behind for certain amount of time because of energy storage in passive components (Inductors and capacitors). The effects on which the distortion in output signal depend on past values of signals are collectively referred as memory effects, Power Amplifier having more memory displays linearity degradation in wide band widths. The memory depth is the period required for sufficient fade away of echoes. The memory effects initiate due to the delay of gain and phase changes in power amplifier corresponding to the frequency and also because of behavior of matching circuits, and transistor capacitances. The higher distortions in Power Amplifier output stimulate bias impedances of the network, resulting low-frequency signals. Echoes inside the bias network passive components are the main cause to modulate the baseband signal, producing third-order intermodulation products.

9 PA Optimization Techniques

The entire efficiency and linearity of a PA design are dictated by its output stage, so careful optimization approaches are required for optimum performance of PA. This section primarily focuses on size, bias point and dimensions of the PA to obtain high efficiency for obtained output power.

9.1 *Parameterized Output Stage*

The layout of unit power cell that is adapted in 1P7M 28 nm CMOS process with cell dimensions $W_{unit}/L_{unit} = 32$ fingers $\times$ 1 μm/28 nm [3]. The scalable nMOS in a push-pull stage which is neutralized [3]. The width of the nMOS, $W_{nMOS} = m \times W_{unit}$, is selected to improve reflection coefficient of the primary stage to achieve high stability. In differential mode PA designs, baluns and LC section tuning circuit are implemented at fundamental frequency. In transformer-based topology, higher odd-order harmonic were terminated to obtain an optimized and reliable simulation values under restricted DOF. This indicates low-order circuits are suitable to design a power amplifier with minimum insertion losses. In order to terminate the high order even-harmonics, switching bypass capacitors are implemented to estimate the ac simulation at input terminal. Insertion losses of impedance matching circuits are estimated to differ with changing values of impedance. Identifying losses in matching circuits for entire model is a challenging task, as we required electromagnetic simulations. Thus, to overcome ambiguity in selection scheme *LC*-tuner loss is deleted. Finally, ground the source and bulk terminals.

9.2 *Optimization Procedure*

By considering V_{GS} and WnMOS as two independent variables, a contour is plotted by superimposing average Pout and the plot is obtained from the following steps at individual WnMOS, V_{GS} from [3]:

(1) Load-Pull: A continuous wave (CW) signal at Pin (VGS) = $Pout_{,req}$-G_{max} (VGS) makes Pout ≈$Pout_{,req}$, where G_{max} (VGS) is the maximum available gain of the

output stage. The estimated application of the required power at output for the duration of load-pull simulation achieves a decline in PAE at later step.

(2) Simulation of EVM: The modulation conversion models like amplitude-to-phase (AM-PM) and amplitude-to-amplitude (AM-AM) modulation at the output stage are obtained from simulation of CW signal.

With examine PAE/EVM output behavior, a design plot is illustrated in figure for EVM versus P_{out} slope at −27 dBc [3]. From plot, it is noticed that EVM value is between 1 and 2 dB when third-order intermodulation IM3 dominates.

9.3 Optimization Results

From the plot of PAE versus VGS-Vt, it is observed a decrease in PAE with increasing VGS-Vt. The contour plot of Pout shows adverse effects, which affect the characteristics of PAE. Considering this complexity, the following limiting factors are noticed, which are explained below.

(1) A horizontal Pout contour is observed for a constant WnMOS, as VGS—Vt increases. These resultant contours specify that Pout is restricted by current clipping at fixed Wn MOS. The current slicing can be avoided by choosing a minimum WnMOS for $P_{sat} \approx P\text{out}_{,req} + \text{PAPR}$, and indirectly satisfy the requirements of EVM. PAE is suboptimal over effective voltage (VGS-Vt) ranging from 100 to 200 mV at which a roughly horizontal Pout contours defend the traditional approach.

(2) For a constant effective voltage (VGS-Vt), the output power contours increase vertically corresponding to increase in WnMOS [3]. Larger value of WnMOS indicates large capacitance at gate and achieves in optimum amplitude-phase conversion, which leads to minimum Pout. In addition, small conduction angle of class-C limits output power through limiting P_{sat} if VGS-Vt is small. A realistic compromise between Pout and PAE of 6.5 dBm is achieved by putting m value as 2 with effective voltage as −150 mV. Moreover, a device technology of $W\text{nMOS} = 384\ \mu\text{m}$ in subthreshold region has nonlinear intrinsic Cgs. Further, quality factor of the device is increased by neutralization of input impedance. Narrow bandwidth of the design is undesired for two reasons. First, the modeling accuracy in cascading amplifiers increases process variable temperature (PVT) sensitivity at Mm-Wave frequency. Second, the excess AM-PM conversion and load modulation at driver stage increase with Pout as the frequency of intermediate stage impedance matching changes to minimum values due to non-linearity in C_{gs}. The above mention inductive degeneration technique in the PA output alleviates the specified case that achieves at a current density value of $10\ \mu\text{A}/\mu\text{m}$ with the help of a reasonably small inductance source. The purpose is to improve output power corresponding to amplitude–amplitude modulation conversion.

10 Design Methodology

The method defined for above targets P_{sat} are alternates for accurate goal designs of Pout and PAE with required EVM. It considers hard slicing actions, while CMOS technology PAs exhibits steady compression. Further, the method failed to highlight the PAE back-off penalty of selecting gate voltage to maximize fmax, and avoid AM-PM. Design approaches involve perfectly optimization of gate width and gate voltage. To obtain optimum design goals, a procedure using an AM-AM or AM-PM modulation schemes for a single-stage PA. Pout and PAE contours of a modulated signal at fixed EVM are traced by considering transistor gate properties as an independent parameters, correspondingly a minimum gate voltage and maximum gate width results in the optimum Pout-PAE negotiate. The resultant nonlinear transistor capacitance enhances AM-PM conversion and proposes a limited memory by intermediate-stage matching bandwidth. The optimized inductive source impedance of a 28-nm process to overcome these challenges with an attainable gain (Table 2).

Besides the merits of using best source impedance, memory still manifested experimentally as the values of EVM decline for bandwidth more than 150 MHz. It is observed that modulation conversion (AM-AM and AM-PM) schemes result zero

Table 2 Performance comparison of state-of-the art 28 GHz CMOS PAs

	ISSCC'17 [2]	ISSCC'16 [1]	RFIC'17-(i) [7]	RFIC'17-(ii) [8]	RFIC'17-(iii) [5]	TMTT'16 [9]
Technology	40 nm CMOS	28 nm CMOS	28 nm CMOS	28 nm CMOS	40 nm CMOS	28 nm CMOS
Frequency (GHz)	27	30	34	32	27	28
Pwr. combining	1	1	2	29 (Doherty)	2	1 2-stack
Supply voltage (V)	1.1	1.0	0.9	1.0	1.0	1.1 2.2
Signal format comp. Carriers PARR (dB)	64 QAM, OFDM 8-CC 9.7	64 QAM, OFDM 1-CC 9.6	64QAM 1-CC 8.3	64QAM 1-CC 8.3	64QAM 1-CC 8.3	64QAM 1-CC 7.5
RFBW (MHz)	800	250	675	2975	667[a]	80
EVM (dBc)	−25	−25	−25	−21.3	−25	−27 −27
Pout @EVM (dBm)	6.7	4.2	8.9	11	8.8	6.8 11
PAE @EVM (%)	11	9	4.4	5.8	9.6	16.5 17

[a]Estimated form data rate and normalized to RMS constellation power

CBW, and henceforth fail to capture the effects of memory. Moreover, two-tone simulations endure memory effects of frequency-based networks. Beginning from the design procedure, dual-resonance transformers that produce double resonance are implemented to achieve wide bandwidths. In addition, redesigned of bias networks that possess lower impedances produce long-term memory. In [3], shows the comprehensive state-of-the-art for 250 MHz, accomplishing a peak value at 800 MHz that are essential for 5G NR.

11 On-Chip Power Combining

One of the modest feasible powers combining technique is shunt connection of CMOS transistors that increase width of the gate, which are subjected to the limitations. The mentioned technique is not profitable as the inductor occupies major on-chip area. In series interconnection of transistors that raises P_{out} by delivering equal VDD for all transistors. This technique increases effective knee voltage, which degrades PAE. Along with the transistor interconnections discussed above, two more power combining techniques like common power combining on on-chip are exhibited in [2]. By combining outputs of two differential devices in a balanced to unbalanced transformer increases the VDD.On-chip magnetic transformers with multiwindings are combined to form two-to-four output voltages of power amplifier. The size of combiner, load impedance mismatches and losses between combined cores are key issues like P_{sat} increases up to 2 to 2.5 dB. In the same way, multiwinding transformers are implemented to couple currents to form multiple cores. The losses in combiner loss are a major concern, which adversely limits the VDD that cannot be improved by using combining techniques.

12 PAE Enhancement Techniques

Conventionally, back-off PAE improvement methods come under high-level design, which targets PA to operate near to Psat. Two techniques that are used by radio frequency PAs are: A Doherty PA, which provides a better linearity whilst able to increase efficiency of the power amplifier when compared with balanced PAs, consists of two different PAs: a carrier PA to achieve high PAE, and a peaking PA which operates to obtain high P_{sat} at peak of the signal. In Doherty amplifiers, the Class AB is generally considered for carrier amplifier, whereas Class C is preferred for peaking amplifier, which conducts for one-half of the full cycle. Both PAs should drive currents of the signal to a load with accurate synchronization, while these mismatches are hard to minimize, significantly affects linearity of the PA [22]. The input signal is divided into two signals with the help of a quadrature coupler. The input behaves similar to a balanced amplifier, where amplifier mismatches are eliminated if the amplitude and phase of the reflection coefficients are equal. Envelope tracking is another technique, which uses an amplifier as envelope to operate as a PA where the signal follows the envelope. The two methods, envelope tracking and out-phasing techniques are unable to meet the 5G data rate requirements. Major challengers are power consumption and also delay mismatches among the paths for out-phasing, and settling time of amplifier.

12.1 Class-G PA with Switching Technology

Envelope tracking is a popular voltage modulation PA technique, which generates minimum back-off power dissipations. This technique improves efficiency at back-off power. Nonetheless, traditional modulation designs usually face a trade-offs accuracy, area, and power, which make CMOS PA implementation a challenging task to meet strict design issues in mobile standards. Power amplifiers Class-G architecture design discusses few major concerns by employing multiple supply voltages, rather than conventional modulation schemes. Hence, a class-G PA controller simply chooses the suitable dynamic switching of supply voltage for discrete envelope tracking. The design of the class-G architecture may further needs huge DC–DC converters.

12.2 Digital Doherty Polar PA with Digitally Reconfigurable Carrier/Peaking Amplifier Paths

In base stations, generally the Doherty PA architecture is widely used to increase PA back-off efficiency. Recently, works on Doherty PAs with CMOS integration prove these designs are optimal candidate for economical mobile applications. Doherty PA obtains a modulation bandwidth and significantly enhanced efficiency at back-off power. Conventional Doherty PAs consist of two analog PAs biased with different conduction angles and the power divider distributes power to both the amplifiers as shown in [4]. However, these traditional designs normally experience degraded performance due to the imperfect matching among two PAs. Particularly, the relative gains between the two PAs match rarely under ideal conditions. Although few analog schemes are proposed, for example, in dynamic biasing the imperfection of two paths leads to design challenges and affect the Doherty PA's performance. Additionally, frequency-dependent components limit the bandwidth of a Doherty PA. The issues in analog techniques are overcome in digital architecture, where it is possible to program gain and phase independently for two paths in a digital fashion as shown in figure and ref from [4]. The gain of the digital PA is measured at both carrier and peaking PAs and the phase controllability is obtained by using a varactor loads. In [4], shows the digital Doherty PA digital architecture implemented in a 65-nm bulk CMOS technology. RF modulated signal is split into two signals with right angle phase shift by a splitter network, and then amplified by respected PA and the obtained outputs are then fed to a PCT combiner network for impedance downscaling. Digital codes are used to control the two DACs for proper real-time synthesize of the modulated signal. To achieve maximum power added efficiency in peaking amplifier, Class D-1 PA topology is employed.

13 Antenna Mismatch Mitigations

In recent developments for 5G cellular communications, the demands for micro-sized antennas and load mismatches in antenna are becoming critical issues. Impedance mismatch tuning and automatic impedance detection are gaining importance in the field wireless communications. Numerous detection techniques are proposed to determine mismatches in antenna and load reflection coefficient. In [4], shows antenna load mismatches. Impedance tuning technique improves PA reliability, since it overcomes device stressing caused by mismatch at load of antenna. The drawback in the mentioned technique is it failed in identifying all load mismatch challenges. Initially, the basic antenna load detection method detects only the amplitude of reflection coefficient but not phase. Moreover, a PA design generally shows decline efficiency at the back-off levels. Also, the impedance mismatch is a time dependent phenomena due to effects of antenna proximity. Hence, an optimum impedance detection and tuning technique should accurately detect the vector impedance with both amplitude and phase information. To analyze vector-based antenna load detection, a feedback architecture power amplifier is implemented as shown in [4]. Later, main PA load impedance is evaluated from V_A and V_j corresponding to the amplitude and phase deviations. Fascinatingly, this architecture automatically tunes to optimal results at minimum output power. If the gain of main amplifiers is reduced in back-off mode by decreasing the gate width, then the load impedance is raised equally by detection and retuning of impedance. Thus, the PA efficiency can be restored to its theoretical optimum value based on the load-line theory. The impedance of the linearly operated main power amplifier is measured by detecting and calibrating the load at increasing ramp stage. A controlling sequencer is implemented by utilizing the capabilities of CMOS technology. The calculation of impedance in closed-loop arrangements is generated by using a burst detector that verifies the PA output power. In order to obtain a antenna load mismatch, an impedance tuner method with minimum loss should be performed through SOI technology. Probably, on-chip integration of an impedance tuner, CMOS-based switches PA can reduce the packaging complexity and form factor. The performance of power amplifier has been implemented for WCDMA applications [4].

14 Pushing Performance Limits: Digital Predistortion

Digital predistortion is a mechanism, which improves the input of PA to compensate its distortions at PA output, which is widely used in single transistor designs. Digital predistortion needs a feedback receiver, which recognizes and manages the output distortion. This strategy primarily aims optimum linearity at circuit level later approach moves to linearization at system-level.

- The signal processing requires necessary corrections for nonlinearities that occur in cascaded PAs. Choose the nontrivial point that identifies the combined response.
- The increase in PAPR and CBW creates stress in the transmitter blocks. A large transmitter DAC is mandatory to hold high value of PAPR. Signal oversampling in the baseband signal processing and Tx DAC are required to obtain broad bandwidth.
- Memory issues are more dominating due to the large CBW, and a tedious nonlinearity representation required to compensate the distortion at PA output.
- The value at peak of CBW in 5G new radio is 5X times to 160 MHz corresponds in WLAN. FB Rx bandwidth of digital predistortion should be 3–5 times of CBW, hence the analog-to-digital converters need adequate number of bits.
- As a result of active changes in antenna impedance, considerable nonlinearity issues arise in PA designs and are required to find out for various power levels and for different beam directions.

References

1. Kumar A Power amplifiers. academia.edu/6002210/Power Amplifiers
2. Shakib S et al (2019) MmWave CMOS power amplifiers for 5G cellular communication. IEEE Commun Mag
3. Shakib S et al (2016) A highly efficient and linear power amplifier for 28-GHz 5G phased array radios in 28-nm CMOS. IEEE J Solid-State Circuits 51(12):3020–3036
4. Wang H et al (2015) The wireless workhorse: mixed-signal power amplifiers leverage digital and analog techniques to enhance large-signal RF operations. IEEE Microw Mag 16(9):36–63
5. Shakib S et al (2017) A wideband 28 GHz power amplifier supporting 8100 MHz carrier aggregation for 5G in 40 nm CMOS. In: IEEE ISSCC digest in technical papers, Feb. 2017, pp 44–45
6. Hashemi H, Raman S (2016) mm-wave silicon power amplifiers and transmitters
7. Zhang Y, Reynaert P (2017) A high-efficiency linear power amplifier for 28GHz mobile communications in 40 nmCMOS. In: IEEE ISSCC digest in technical papers, June 2017, pp 33–36
8. Raghavan V et al (2017) Millimeter wave channel measurements and implications for PHY layer design. IEEE Trans Antennas Propag 65(12):6521–6533
9. Li XJ, Zhang YP (2010) Flipping the CMOS switch. IEEE Microw Mag 11(1):86–96

Techniques to Improve Gain-Bandwidth 5G ICs

R. Vignesh, Rajesh Kumar, Hanjung Song, and Sandeep Kumar

Abstract This chapter introduces a basics of designs and techniques to improve gain-bandwidth for 5G ICs. The major focus would be on the various network topologies that yield to provide easy implementation of on-chip components for 5G-ICs. Section 1 discusses the basics of RLC tank networks, which includes RC parallel network, RLC network and series to parallel resonant network. The parameters such as quality factor, noise of filter networks are shortly refresh while foundation of resonant circuits would set-up for 5G transceiver ICs. Section 2 introduces coupled resonator networks can be used as microwave components to achieve a better gain-bandwidth trade-off. Finally, Sect. 3 will provide transformer resonators and circuit to reduce bulky components and enhance gain-bandwidth of ICs.

1 RLC Tank

In tank circuit, RL, RC and RLC combinations are popular choices for improving the performance parameters of the RFIC design. The gain-bandwidth trade-off for RL and RC networks limits to impedance matching due to incapable of degree of freedoms.

R. Vignesh · S. Kumar (✉)
Department of Electronics and Communication Engineering, National Institute of Technology Surathkal, Surathkal, Karnataka, India
e-mail: sandeep@nitk.edu.in

R. Kumar
Indian Institute of Technology Dhanbad, Dhanbad, Jharkhand, India

H. Song
Department of Nano Science and Engineering, Center for Nano Manufacturing, Inje University, Gimhae 621-749, Republic of Korea

S. Singh et al. (eds.), *CMOS Analog IC Design for 5G and Beyond*, Lecture Notes in Electrical Engineering 719, https://doi.org/10.1007/978-981-15-9865-4_8

1.1 RC Parallel Network

The circuit admittance from the basic RC parallel network can be expressed as [1].

$$\mathrm{Y} = \mathrm{sC} + \frac{1}{\mathrm{R}} = \frac{sCR + 1}{R} \tag{1}$$

The impedance can simply become $Z = \frac{1}{Y}$ and display a low-pass curve. The resistor of network is responsible for the thermal noise. The output noise current is calculated as

$$\mathrm{I}_{\mathrm{N}}^{-2} = 4\mathrm{KBT/R} \tag{2}$$

where K is Boltzmann's constant, B is bandwidth, T is temperature, R is normalized resistance with reasonable assumption. The noise produces an output noise voltage of $V_{n,out}^{-2}$ which derived by using filter transfer function. At the higher value of filter capacitance, the total integrated noise could be K_B *T/C*. The filter quality factor can be expressed as

$$\mathrm{Q} = \frac{Im\{Y\}}{Re\{Y\}} = \frac{Im\{Z\}}{Re\{Z\}} = \omega RC \tag{3}$$

1.2 RLC Network

In RLC network, combination of R, L and C can provide a degree of freedoms for RF resonance behavior. The low-pass RC filter is converted to $\omega_0 = 1/(\sqrt{}\mathrm{LC})$ by using an inductor. The schematic of RLC network and its output noise performance is given in [1]. In general, inductor and capacitor resonate at f_0 where the tank reduces to be $\mathrm{R_T}$.The output noise voltage of RLC network can be expressed as.

$$V_N^{-2}(\omega) = \frac{4KBT}{R_T} R_T^2 = 4KBTR_T \tag{4}$$

where $\mathrm{R_T}$ is total equivalent resistance,ω_0 is resonating frequency. The quality factor of the filter is

$$\mathrm{Q} = \frac{Im\{Y\}}{Re\{Y\}} = \frac{Im\{Z\}}{Re\{Z\}} = \omega_0 RC \tag{5}$$

The modeling of RLC network can have a simple example of great importance in integrated circuits (ICs). The oscillator in steady-state condition acts as a current source in parallel with RLC tank could effectively modeled [2, 3]. Therefore, it is expectable reason that tank circuit with high quality factor could provide benefit in terms of noise. Similarly, an RF amplifier using parallel RC output impedance can modeled as a voltage driven current source.

1.3 Series-Parallel Resonant Network

In [4], a parallel to series resonant network is used as a bandpass filter. This resonant network achieved a bandpass response and dual band response by selecting proper values of inductors and capacitors. For the parallel and series resonance networks, the impedances Z_p and Z_s are shown in Eq. (6) and (7) respectively.

$$Z_p = \left(\frac{1}{R_1} + jwC_1 + \frac{1}{jwL_1}\right)^{-1} \tag{6}$$

$$Z_s = R_2 + jwL_2 - j\frac{1}{wC_2} \tag{7}$$

Here, the concept implies that different values of source and load impedances $R_1 = 250\Omega$ and $R_2 = 40\Omega$ in parallel to series resonant network offered dual band response with a dip in the middle. A wide band response is achieved when parallel-series network is used as inter-stage matching network between CG and CS LNA. However parallel-series resonant network is inverted as series-parallel resonance network, which is used as matching network between CG-CS LNA for improving performance parameters like impedance matching, noise figure and gain over wider range of operating frequencies. The schematic of two stages CG-CS LNA using series-parallel resonant network. Selection of on-chip inductors is the key for the designing of a matching network, as it determines the performance quality of matching network. The input and inter-stage matching network is accomplished with the combination of a series-parallel LC impedance transformation network. In order to maintain the 50Ω impedance matching, first a single stage of CG LNA using series to parallel resonant network as an input is designed that provides good reverse isolation parameter over a wide band range of operation. It is well known that the NF and high gain are the essential requirements of the LNA design. For improving these parameters further, the next stage, i.e., CS with source degeneration via series–parallel resonant network is combined. For the two-stage of CG-CS LNA, the width of M_1 (WM_1) is calculated to be 492 μm using Eq. (77) as given in [20]. The second stage of CS device is designed to meet the desired performance criterion while width of M_2 (WM_2) is calculated as 532 μm using Eq. (9) as given in [20].

$$\mathrm{WM}_1 = \frac{3}{2}\frac{c_{gs}}{L_{min}c_{ox}} \tag{8}$$

$$WM_2 = \frac{g_m^2 L_{min}}{2K_n I_D} \tag{9}$$

where, L is channel length, K_n is transconductance parameters, c_{ox} is gate oxide capacitance, I_D is the drain current, g_m is transconductance of device and c_{gs} is the gate to source capacitance. The designed circuit of CG-CS LNA is simulated and obtained reflection coefficient of −21.4 and −21 dB at operating frequency of 20.5 and 26.2 GHz, respectively. This return loss shows a frequency band from 19.1 to 28.8 GHz with a fractional bandwidth of 40.5%. It also shows forward gain S_{21} of the circuit in the range of 7.4–21.3 dB for the entire band of operation. The forward gain (S_{21}) shows a high gain over the bandwidth of 9.7 GHz ranging from 19.1 to 28.8 GHz.

1.4 Technique to Extract RLC Network

The equivalent model of rectangular slotted patch antenna, which is extracted with the help of electromagnetic (EM) software package in ADS, consists of two capacitors (C_1 and C_2) that are connected in parallel with an inductor (L) resulting in an L–C circuit. The source of 50Ω in L–C network is represented by coaxial feed probe. To better understand the extracted model, a rectangular slot patch antenna is further divided into individual C circuit and L circuit, respectively. The approximate equivalent circuit of extracted L-C model. From these individual circuits, R_s' for C and L circuit is given as in Eq. (10) and (11) respectively [6].

$$R'_s = R_s\left(\frac{C_1}{C_2}\right)^2 \tag{10}$$

$$R'_s = R_s\left(\frac{m}{m_1}\right)^2 \tag{11}$$

The equivalent capacitance $\left(CT'\right)$ will resonate with inductor in parallel with C_1 in series with C_2 and shown in Eq. (12).

$$CT' = \frac{C_1 C_2}{C_1 + C_2} \tag{12}$$

The calculated values of components for equivalent model achieve wide bandwidth of around 9.7 GHz within the band range of 10.7 to 19 GHz.

2 Coupled Resonator Circuits

The coupled resonator circuit is important components for design of RF and microwave integrated circuits. Despite the different physical structures at hand with different types of resonators, a general technique of coupled-resonator filters is used. However, these design methods can be used to the design of waveguide filters, dielectric resonator filters and microstrip filters. The general coupling matrix can be used to represent a coupled resonator filter.

2.1 *Methods of Coupling Matrix*

The n-coupled resonator is shown in [7]. i is the loop current and e_s denotes the voltage source. The two-port equivalent circuit is shown in [7]. Using the mesh equations and nodal voltages, the normalized impedance and admittance matrix have been derived. The final normalized impedance $[\bar{z}]$ and admittance matrix $[\bar{y}]$ can be depicted in Eqs. 13 and 14 [7].

$$[\bar{z}] = \begin{bmatrix} \frac{1}{q_{e1}} + p - jm_{11} & -jm_{12} & \cdots & -jm_{1n} \\ -jm_{21} & -jm_{22} & \cdots & -jm_{2n} \\ \vdots & \vdots & \vdots & \vdots \\ -jm_{n1} & -jm_{n2} & \cdots & \frac{1}{q_{en}} + p - jm_{nn} \end{bmatrix} \tag{13}$$

$$[\bar{y}] = \begin{bmatrix} \frac{1}{q_{e1}} + p - jm_{11} & -jm_{12} & \cdots & -jm_{1n} \\ -jm_{21} & -jm_{22} & \cdots & -jm_{2n} \\ \vdots & \vdots & \vdots & \vdots \\ -jm_{n1} & -jm_{n2} & \cdots & \frac{1}{q_{en}} + p - jm_{nn} \end{bmatrix} \tag{14}$$

where q_{e1} and q_{en} are external quality factors in scaled form, m_{ij} denotes the normalized coupling coefficients and p denotes the complex frequency variable. The normalized impedance matrix $[\bar{z}]$ is equal to the normalized admittance matrix $[\bar{y}]$. It is much essential because it could have a unique formulation for an n-coupled resonator filter. So, the generalized S parameter can be depicted as shown in Eq. 15 and 16.

$$S_{21} = 2\frac{1}{\sqrt{q_{e1} \cdot q_{en}}}[A]_{n1}^{-1} \tag{15}$$

$$S_{11} = \pm\left(1 - \frac{2}{q_{e1}}[A]_{11}^{-1}\right) \tag{16}$$

with

$$[A] = [q] + p[U] - j[m]$$

where $[U]$ is the n × n identity matrix, $[q]$ is an n × n matrix for all entries zero. $[m]$ is general coupling matrix. In general coupling matrix $[m]$, the synthesis procedure will have all nonzero values. For an asynchronously tuned filter, the nonzero value will occur only in the diagonal elements. The nonzero means that the coupling exists between every resonator with every other resonator that the network $[m]$ represents. In order to perform a sequence of transformations until a more convenient method have to obtain for implementation.

3 General Theory of Couplings

The next step in designing the filter is to develop a relationship between the coupling coefficient and the coupled resonators. Therefore, it is easy to find the physical dimensions of the filter toward the fabrication of structure. The coupling coefficient "k" can be defined as the ratio of coupled energy to the stored energy. The coupling coefficient varies on the basis of their structure and having different self-resonating frequencies. k is defined as mathematically in Eq. (17) [7].

$$k = \frac{\iiint \varepsilon \underline{E}_1 \cdot \underline{E}_2 \, dv}{\sqrt{\iiint \varepsilon \left|\underline{E}_1\right|^2 dv \times \iiint \varepsilon \left|\underline{E}_2\right|^2 dv}} + \frac{\iiint \mu \underline{H}_1 \cdot \underline{H}_2 \, dv}{\sqrt{\iiint \mu \left|\underline{H}_1\right|^2 dv \times \iiint \mu \left|\underline{H}_2\right|^2 dv}} \tag{17}$$

It is defined as the volume integral over all affected regions with permittivity of ε and permeability of μ. In [7], general coupled RF and microwave resonators are given. In the coupling coefficient equation, the first term represents the electric coupling and the other term represents magnetic coupling. The coupling fields either have a positive or negative sign. A positive field signifies that a coupling could increase the stored energy of uncoupled resonators while a negative field demonstrates the reduction. So, the same sign indicates that magnetic and electric coupling is having the same effect whereas the opposite sign indicates they have the opposite effect. The knowledge of the field distribution is required to solve the Eq. (17). However, it is tough task unless analytical solutions of the fields exist. In order to find some characteristic frequencies which are associated with coupled RF and microwave resonators may be much easier on the other side using the full-wave EM simulation or experiment.

4 Synchronously Tuned Coupled Resonator Circuits

4.1 Electric Coupling

In [7], depicted an equivalent model of lumped element for electrically coupled microwave resonators. The circuit equivalence of lumped element can be narrow band near its resonance if the coupled structure chosen as distributive element. From figure ref [7], the two-port network can be equated as.

$$\begin{aligned} I_1 &= j\omega C V_1 - j\omega C_m V_2 \\ I_2 &= j\omega C V_2 - j\omega \end{aligned} \tag{18}$$

From Eq. 18, the Y-parameters are written in Eq. (19) and C_m represent the mutual capacitance.

$$\begin{aligned} Y_{11} &= Y_{22} = j\omega C \\ Y_{12} &= Y_{21} = -j\omega C_m \end{aligned} \tag{19}$$

The alternate form of the equivalent circuit is shown in [7]. This form is more convenient for our discussion. The admittance inverter is calculated as $J = \omega C_m$. Once the electric wall is inserted then coupling effect enhances the capability to store charge. While when the magnetic wall is inserted the coupling effect reduces store charge of a single resonator. The coupling coefficient k_E can be found using above equations.

$$k_E = \frac{f_m^2 - f_e^2}{f_m^2 + f_e^2} = \frac{C_m}{C} \tag{20}$$

This electric coupling coefficient emphasizes the general definition, which corresponds to the ratio of coupled electric energy to the stored energy.

4.2 Magnetic Coupling

In [7], illustrates an equivalent model of lumped element. From figure ref [7], the two-port network equation can be expressed as.

$$\begin{aligned} V_1 &= j\omega L I_1 + j\omega L_m I_2 \\ V_2 &= j\omega L I_2 + j\omega L_m I_1 \end{aligned} \tag{21}$$

From Eq. 21, the Z parameters are written in Eq. (22) and L_m represent the mutual inductance.

$$\begin{aligned} Z_{11} &= Z_{22} = j\omega L_m \\ Z_{12} &= Z_{21} = j\omega L_m \end{aligned} \tag{22}$$

The impedance inverter $K = \omega L_m$ shows the magnetic coupling. To find the resonant frequency replace the symmetry plane by a short circuit or electric wall or by an open circuit or magnetic wall and find the resonant frequency can be expressed as.

$$f_e = \frac{1}{2\pi\sqrt{C(L - L_m)}} \tag{23}$$

$$f_m = \frac{1}{2\pi\sqrt{C(L + L_m)}} \tag{24}$$

By noticing the above Eqs. (23) and (24), it is noted that when the electric wall is inserted coupling effect reduces the capability to store flux of a single resonator and when the magnetic wall is inserted the coupling effect enhances the capability to store flux. The magnetic coupling coefficient k_M can be found.

$$k_M = \frac{f_e^2 - f_m^2}{f_e^2 + f_m^2} = \frac{L_m}{L} \tag{25}$$

The magnetic coupling coefficient emphasizes the general definition, which corresponds to the ratio of coupled magnetic energy to the stored.

4.3 Mixed Coupling

In [7], shows a network of coupled resonator circuit. The Y and Z parameters can be expressed as:

$$\begin{aligned} Y_{11} &= Y_{22} = j\omega C \\ Y_{12} &= Y_{21} = -j\omega C'm \\ Z_{11} &= Z_{22} = j\omega L \\ Z_{12} &= Z_{21} = j\omega L'm \end{aligned} \tag{26}$$

where C is the self-capacitance, L is the self-inductance and $C'm$ and $L'm$ are the mutual capacitance and inductance. The impedance inverter $K = \omega L'm$ indicates the magnetic coupling. While admittance inverter $J = \omega C'm$ indicates the electric coupling.

If we assume $L'_m C'_m << LC$ the Eq. 31 becomes

$$k_X \approx \frac{L'_m}{L} + \frac{C'_m}{C} = k'_M + k'_E \tag{27}$$

From the Eq. (27), it is clear at the point of observation that superposition of the magnetic and electric couplings results into mixed coupling. These two coupling are extensively used to improve gain-bandwidth product in the 5G ICs.

5 Transformer-Based Resonators

In RFIC, transformers are most popular choice for integrated circuit such as voltage-controlled oscillators, mixers, power amplifiers and low-noise amplifiers for power combining, or impedance conversion. In [8–14], on-chip transformers exhibit low quality factors and could produce low coupling values of coefficients.

6 Inverting and Noninverting Transformer

From figures from ref [15], shows the 3D picture of typical inverting and noninverting transformers in standard 1:1 ratio while their corresponding equivalent lumped element models [16]. The parasitic to the silicon substrate C_{ox}, C_{Si},r_{Si}, the parasitic intra-winding capacitance C_{m1}, C_{m2} and the inter-winding capacitance C_c are shown in [15]. They are particularly based on the analytical and fundamental approach. However, it really a challenging task to develop a flexible network to obtain on-chip transformer [17]. Therefore, authors is need to be formulated these networks based on electromagnetic (EM) RF simulators and to provide accurately modeling of the network [18]. Furthermore, it is instructive to focus on the equivalent model of simplified differential mode (DM) transformer coupled network. The intrawinding capacitance can be modeled as an equivalent parallel RC network. In order to further simplify, the effect of the parasitic of interwinding capacitance C_C, following R_{S1} and R_{S2} is neglected. This resulting into two-port filter, schematic, is shown in [15]. If the current flowing across interwinding capacitance, i.e., $I_{CC} = 0$ then it is expected that C_C has not provided any effect. Similarly, the effect of C_C can also be increased when I_{CC} is kept maximum. It is interested that the voltage across C_C can be written as:

$$V_{CC} = V_1 - V_2 = I_1(|Z_{21}|e^{j\angle Z_{21}}\left(\frac{Z_{21}}{Z_{11}}\right)e^{j(\angle Z_{11} - \angle Z_{21})} \tag{28}$$

where V_1 is input voltage, V_2 is input voltage and I_1 is the input current. The equation shows that voltage drop across C_C is maximum when $\angle Z_{11} - \angle Z_{12} = \pm 180°$ and minimum when $\angle Z_{11} - \angle Z_{12} = 0°$ regardless the magnitude of Z_{11} and Z_{12}. Furthermore, investigate the parasitic effect, the previous design example will be refreshed so let us assume an inter-winding capacitance, i.e., $C_C = 1\text{fF}$. This value is reasonable for center frequency of $f_O \approx 80$ GHz if low-k transformers are used and it is more dominant when high-k transformers are considered. If C_C is neglected then magnetic coupling coefficient k has no effect is appeared on the magnitude of the trans-impedance of the filter, and it resulting into 3-dB B.W ≈ 31.3 GHz along with approximately of 0.16 dB in-band ripple [1]. However, it is observed that inverting and noninverting transformers behave as very distinctly when the parasitic inter-winding capacitance is considered. An inverting transformer ($k < 0$) recognizes $\angle Z_{11} - \angle Z_{12}$ that come out from the proximity of the low frequency pole f_L to $\approx -10°$ while the proximity of the high frequency pole f_H [5]. Consequently, the C_C lowers f_L while keeping the value of f_H is high. In [16], illustrated a wider bandwidth and larger ripple. In contrast, the same filter uses a noninverting transformer for $k > 0$ and can be seen in figure available [16]. In this second case, f_H is fall out at lower frequencies while f_L does not alter. This results into lower bandwidth of 9 GHz. It is an important note for broadband designs to counter-act the effect of C_C, a lower L_1, L_2 and k can be used when a transformer with $k < 0$ is adopted. This provides a further reduction of the parasitic inter-winding capacitance. However, in order to counter act the effect of C_C, a larger value of k is needed when a noninverting transformer is used. This provides a further enhancement of the parasitic interwinding capacitance. For this reason, it is more desirable to go toward inverting transformers for achieving the better networks whenever possible [9]. It is finally assuming as $L_1 = L_2 = L$, the self-resonant frequency can be derived as:

$$f_{SRF} = \frac{1}{2\pi\sqrt{2L(1-K)C_c}} \tag{29}$$

For inverting (f_{SRF}, $k < 0$) and noninverting transformer (f_{SRF}, $k > 0$) with same L, C_c and $|k|$, and then the self-resonant frequencies can be expressed as:

$$f_{SRF} = \sqrt{\frac{(1-|K|)}{(1+|K|)}}(f_{SRF}, k > 0) \tag{30}$$

It noting that, f_{SRF} $k > 0$ is always greater than f_{SRF}, $k < 0$. Because $k > 0$ is always preferable over high frequency of operation. If assuming $C_1 = C_2 = C$, then the expressions of ω_L and ω_H can be derived. The interesting point to observe that different filters are reported and responded to this effect. The same approach as followed and carried out for frequency of the complex poles when $C_1 = nC_2$ and that can be expressed as in Eqs. (31) and (32) respectively [19–21].

$$\omega_L = \frac{1}{2\pi\sqrt{2L_{C1}(C_1 + C_c\left(1 + {}^1\!/_n\right)}} = \frac{1}{\sqrt{2L_{L1}C_1}} = \frac{1}{\sqrt{2L_{L2}C_2}} \tag{31}$$

$$\omega_H = \frac{1}{2\pi\sqrt{2L_{C1}C_1}} = \frac{1}{\sqrt{2L_{L2}C_2}} = \frac{1}{\sqrt{2L_{L2}C_2}} = \frac{1}{\sqrt{2L_{M2}(1-|K|)C_2}} \tag{32}$$

Filter is not able to follow the simple design guidelines or insight. In [15], shows the frequency response of the filters with effective value of $n = 2$. The typical value of $n = 2$ is effective for the inter-stage matching network. However, it is an extreme case for an LNA where the size of the amplifiers is unable to increase the saving power [22, 23]. The two resonance frequencies ω_L and ω_H can be illustrated as in Eqs. (33) and (34).

$$\xi = \frac{\mathrm{L_{M2}C_2}}{\mathrm{L_{M1}C_2}} \tag{33}$$

when $\xi = 1$, the complex poles can be expressed as [24, 25].

$$\omega_{L,H}^2 = \frac{1 + \xi + \sqrt{(1+\xi)^2 - 4\xi\left(1 - K_M^2\right)}}{2\mathrm{L_{M2}C_2}\left(1 - K_M^2\right)} \tag{34}$$

It is interestingly that the magnetically coupled resonators are not affected by their frequency responses, except reduction in transimpedance gain for a 10log10 (n) curve. It is not the case of other 4th order filters. To overcome this issue, inductively coupled resonators offer a design method that applies Norton transformation to derive a transformer-based filter. The design parameters are derived in a one-step from the Eqs. (33) and (34), which is end result. Finally, both magnetically and capacitive coupled resonators show a frequency response if condition $C_{MC} = -Ck/(1 + kCQ_{CMC})$. This demonstration clearly shows the robustness of the above design techniques.

7 Conclusion

In this chapter, reported architectures such as RLC tank circuits, coupled resonators circuit and transformer coupled circuits have been discussed. These architectures are derived fundamentally and showed their performance to improve gain-bandwidth of 5G ICs.

References

1. Vigilante M, Reynaert P (2018) 5G and E-band communication circuits in deep-scaled CMOS, analog circuits and signal processing. Springer Nature, New York
2. Fano RM (1950) Theoretical limitations on the broadband matching of arbitrary impedances. J Franklinst 249(1):57–83

3. Bevilacqua A, Niknejad AM (2004) An ultra-wide band CMOS low-noise amplifier for 3.1–10.6-GHz wireless receivers. IEEE J Solid-State Circuits 39(12):2259-2268
4. Lo YT, Kiang JF (2011) Design of wideband LNAs using parallel to series resonant matching network between common-Gate and common-source stage. IEEE Trans Microw Theory Techniq 59(9):2285–2293
5. Kumar S, Kumar B, Kanaujia SD (2015) Co-Design approach for wide-band asymmetric cross shaped slotted patch antenna with LNA. Wirel Pers Commun (Springer) 85(3):863–877
6. Kumar S, Kumar B, Kanaujia SD (2015) Co-Design and analysis of low noise optimization amplifier using reconfigurable slotted patch antenna. Wirel Pers Commun (Springer) 97(4):5185–5200
7. Hong JS, Lancaster MJ (2001) Microstrip filters for RF/microwave applications. Wiley. 0-471-22161-9 (Electronic)
8. Yang J, Kim CY, Kim DW, Hong S (2010) Design of a 24-GHz CMOS VCO with an asymmetric-width transformer. IEEE Trans Circuits Syst II, Exp Brief 57(3):173–177
9. Pye AD, Hella MM (2011) Analysis and optimization of transformer based series power combining for reconfigurable power amplifiers. IEEE Trans Circuits Syst I, Reg Pap 58(1):37–50
10. Chen HC, Wang T, Chiu HW, Yang YC, Kao TH, Huang GW, Lu SS (2009) A 5-GHz-band CMOS receiver with low LO self-mixing front end," IEEE Trans Circuits Syst I, Reg Pap 56(4):705–713
11. Im D, Nam I, Kim H-T, Lee K (2009) A wideband CMOS low noise amplifier employing noise and IM2 distortion cancellation for a digital TV tuner. IEEE J Solid-State Circuits 44(3):686–696
12. Han J, Choi B, Seo M, Yun J, Lee D, Kim T, Eo Y, Park SM (2010) A 20-Gb/s transformer-based current-mode optical receiver in 0.13- μm CMOS. IEEE Trans Circuits Syst II, Exp Brief 57(5):348–352
13. Carrara F, Italia A, Ragonese E, Palmisano G (2006) Design methodology for the optimization of transformer-loaded RF circuits. IEEE Trans Circuits Syst I, Reg Pap 53(4):761–768
14. Long JR (2000) Monolithic transformers for silicon RF IC design. IEEE J Solid-State Circuits 35(9):1368–1382
15. Vigilante M, Reynaert P (2017) On the design of wideband transformer-based fourth order matching networks for e-band receivers in 28-nm CMOS. IEEE J Solid-State Circuits 52(8):2071–2082
16. Decanis U, Ghilioni A, Monaco E, Mazzanti A, Svelto F (2011) A low-noise quadrature VCO based on magnetically coupled resonators and a wideband frequency divider at millimeter waves. IEEE J Solid-State Circuits 46(12):2943–2955
17. Aoki I, Kee SD, Rutledge DB, Hajimiri A (2002) Distributed active transformer-a new power combining and impedance-transformation technique. IEEE Trans Microw Theory Techol 50(1):316–331
18. Aoki I, Kee SD, Rutledge DB, Hajimiri A (2002) Fully integrated CMOS power amplifier design using the distributed active-transformer architecture. IEEE J Solid-State Circuits 37(3):371–383
19. Haldi P, ChowdhuryD, Reynaert P, Liu G, Niknejad AM (2008) A 5.8 GHz 1 V linear power amplifier using a novel on-chip transformer power combiner in standard 90 nm CMOS. IEEE J Solid-State Circuits 43(5):1054–1063
20. Kaymaksut E, Franois B, Reynaert P (2013) Analysis and optimization of transformer-based power combining for back-off efficiency enhancement. IEEE Trans Circuits Syst I: Reg Pap 60(4):825–835
21. Analui B, Hajimiri A (2004) Bandwidth enhancement for transimpedance amplifiers. IEEE J Solid-State Circuits 39(8):1263–1270
22. Wu CH, Lee CH, Liu SI (2005) CMOS wideband amplifiers using multiple inductive-series peaking technique. IEEE J Solid-State Circuits 40(2):548–552
23. Kaymaksut E, Reynaert P (2015) Dual-mode CMOS Doherty LTE power amplifier with symmetric hybrid transformer. IEEE J Solid-State Circuits 50(9):1974–1987

24. Ozen M, Anderson K, Fager C (2016) Symmetrical Doherty power amplifier with extended efficiency range. IEEE Trans Microw Theory Technol 64(4):1273–1284
25. Chappidi CR, Sengupta K (2016) 20.2 a frequency-reconfigurable Mm-Wave power amplifier with active-impedance synthesis in an asymmetrical non-isolated combiner. In: IEEE International solid-state circuits conference (ISSCC). San Francisco, CA, pp 344–345

GaN-Based Technology for 5G Applications

Brajendra Singh Sengar, Amitesh Kumar, Maddaka Reddeppa, Saurav Kumar, and Htet Ne Oo

Abstract This chapter presents the GaN-based different devices for 5G applications. It includes the design and performance optimization of devices for the above mentioned application. The comparison between Si-based devices with GaN-based devices has been performed and The performance of GaN-based devices outplayed the Si-based devices.

1 Introduction

As we know, the GaN-based technology for RF and microwave application have been industrially available for several years now and continue to advance. As shown in Table 1, primarily GaN has been developed for other circuit applications [1]. However, currently, GaN has been seen as the next-generation power amplifier (PA) technology [1–4]. The present channel length range varies from 0.10 μm to 0.15 μm in GaN-based HEMT [1]. There are two important areas of communication (Satellite and Cellular) that can enhance GaN industry.

Both Authors have worked together on this publication and contributed equally.

B. S. Sengar (✉)
Department of Energy Science and Technology, Centre for Advanced Studies, Lucknow, India
e-mail: brajendra@cas.res.in

A. Kumar (✉)
Department of Electrical Engineering, National Institute of Technology, Patna, India
e-mail: amitesh.ee@nitp.ac.in

M. Reddeppa
Department of Physics, Chungnam National University, 99 Daehak-Ro, Yuseong-gu, Daejeon 34134, Republic of Korea

M. Reddeppa · S. Kumar
CSIR-Central Scientific Instrument Organization, Chandigarh, India

H. N. Oo
Department of Information Science (Faculty of ICT), University of Technology, (Yatanarpon Cyber City), Pyinoo Lwin, Myanmar

S. Singh et al. (eds.), *CMOS Analog IC Design for 5G and Beyond*, Lecture Notes in Electrical Engineering 719, https://doi.org/10.1007/978-981-15-9865-4_9

Table 1 GaN commercially available GaN devices [5]

Foundry Service			Discretes
Process	Bias (V)	Freq (GHz)	
0.25 μm GaN-on-SiC	28–40	18	Y
0.40 μm GaN-on-SiC	28–50	8	Y
0.25 μm GaN-on-Si	N/A	N/A	N
0.50 μm GaN-on-Si	N/A	N/A	N
0.50 μm GaN-on-SiC	N/A	N/A	N
0.20 μm GaN 4-in	N/A	60	Y
0.10 μm	N/A	70	N
0.25 μm GaN-on-SiC 100 mm	40	18	Y
0.25 μm GaN-on-SiC 100 mm	48	10	Y
0.15 μm GaN-on-SiC 100 mm	28	40	Y
0.50 μm GaN-on-SiC 100 mm	65	10	Y
0.50 μm GaN-on-SiC 3-in E-mode	N/A	N/A	N
0.15 μm GaN-on-SiC 3-in	N/A	Ka-band	N
0.50 μm GaN-on-SiC 3-in	40 V	X-band	N
0.25 μm GaN-on-SiC	N/A	30	Y

In this chapter, we have discussed the urgency of GaN-based RF technology. Data showing GaN's potential development for mainly RF and microwave applications have been discussed. An output stage for the 5G PA technology can be designed with high power added efficiency (PAE) using a single transistor only. Linearity, low conversion loss and better dynamic range can be achieved by utilizing a satellite communication mixer.

2 Gallium Nitride Potential Technology for 5G

From the perspective of Power IC designers, they want high frequency, and highly efficient devices because it will reduce the power consumption and heat losses during signal transmission or energy conversion. [6, 7]. At the higher operation frequency, the conversion loss in the system and the magnetic energy storage volume can be highly reduced [7]. Moreover, using monolithically integrated circuits, the parasitic inductance induced by bonding wires, diodes, transistors, and other passive components can be severely reduced. It will again enhance the frequency and efficiency. This increase in frequency and efficiency has benefiting the power integrated circuits industry [8, 9].

Apart from the power IC design development techniques, the performance of power integrated devices has been improving continuously. The continuous improvement in the performance of devices has enhanced the development and search of novel materials [10, 11]. In comparison to Si, GaN material is having more than 2.5 times of electron saturation velocity, 3 times of energy bandgap, and 11 times of critical electric field. These properties make GaN one of the best materials for RF and power conversion applications [12]. To fully utilize the merits of GaN material, a 2-Dimensional Electron Gas (2DEG) channel-based device such as High Electron Mobility Transistors (HEMTs) exhibit and well fit the lateral integration [13].

In 1975, the first GaAs-based Monolithic Microwave Integrated Circuit (MMIC) has been proposed [14]. Till now, In RF and microwave applications of GaN-based devices, MMIC is the conventional technology [7, 15]. For MHz to GHz ranges, HEMT devices with single-chip air-bridge combination, epitaxy resistors or thin-film, and metal-insulator-metal (MIM) capacitors, MMIC has been used for the high-frequency microwave electronics [16, 17]. While the development progresses, the dimension of GaN-based MMIC technology has been scaled down from the microns scale to submicron scale. These are some of the submicron number (0.5 μm [18], 0.25 μm [19], 0.15 μm [20], 0.1 μm [21], 0.04 μm and 0.02 μm [22]) reported by different research groups around the world.

Industrial HEMT GaN-based MMIC technology includes Fujitsu, BAE Systems, Fraunhofer, NXP, Oki, MACOM, and HRL. For 5G mobile bands, the GaN HEMTs have been excellent MMIC power amplifiers with higher PAE performance. Along with the RF and microwave applications, power conversion-based applications can also use the MMIC process for better performance. DC–DC is generally working around 1 GHz. However, GaN-based MMIC works around 100 MHz because it has relatively short length of the drift area [23–25]. Furthermore, HEMT is having a negative threshold voltage due to this HEMTs can have faults and involves gate driver, which is designed especially for these applications. These faulty conditions hamper GaN MMIC developments [26, 27]. Normally, off operation of GaN power HEMTs are introduced through p-type cap layer [27], gate recess [28], implantation using fluorine-ion [28]. A new technology of high-frequency and power conversion high-efficiency has been introduced by GaN normally OFF power devices. Companies like Infineon, Texas Instruments, Cree, Panasonic, ON-semi have launched their GaN products. As the industrial production of these devices increases, the fabricated GaN-based power convertor (normally-off transistors) has outplaced the Si-based transistors because GaN-based transistor is having higher output efficiency and operation frequency.

3 GaN for 5G

The infrastructure of cellular communication is getting complex as technology moving forward from 2G-based communication to LTE and now 5G technology. The 5G technology is 1000 times faster than that of LTE. It is also changing the existing telecom infrastructures for emerging technologies such as self-driving cars, wearable technology, the internet of things (IoT), and modern medical implantable devices [29]. As we know 5G technology will be enhancing the features of IoT so it needs to have high speed, highly dense, and very low latency.

For higher over-the-air data rates, The MIMO and beamforming technology are required to direct signal power in the 5G technology. 5G systems have achieved more than 3 Gb/s, several demonstrations have been performed by different groups [30]. Dimension and efficiency of the RF transceiver hardware can be further redesigned for better performance using these MIMO architectures. Numerous new ideas for 5G technology has been emerging. Primarily, the design and execution of transmitter–receiver hardware will be defined by the spectrum allocation. As it is known that the 6-GHz cellular communication bands are highly congested. Therefore, it is very essential to use mm frequencies for realizing the anticipated low-latency, high-speed communication. Several bands for the dominant cellular communication including 28 GHz (Verizon), 39 GHz (T-Mobile), and 37 GHz have been approved by the FCC [31].

Array of RF circuitry is needed for every antenna division in the transceiver system of the phased array. For this purpose, massive MIMO beamforming can be used. Consequently, the important figure of merit for the design of handset is size, cost, and power density of communication. Similar figure of merit applied for base station as well. Different types of beamforming techniques like analog, digital, and hybrid have been in discussion. An array of RF transmitters and receivers will be needed. Although there are technologies that are superior for handsets design because of cost-effectiveness, voltage efficiency battery, and requirements of RF power, GaN is a accepted contender for base-station arrangement. There are a lot of efforts have been made to utilize GaN for high frequency communication and low operating voltages. This will improve the advancement of switches and other components. GaN-based semiconductor devices are having better performance because it has many advantages over other solid-state materials. That is the reason they are used for high power applications (microwave, RF, and power conversion).

The GaN material is having a wurtzite structure. The electronegativity value of Ga and N atoms is different. Due to this, the cores of the positive and negative charge of the electron cloud do not collide. Therefore, GaN displays spontaneous polarization. For the epitaxially deposited structure of AlGaN/GaN, as we know there will be some lattice mismatch between GaN and AlGaN. This lattice mismatch will develop piezoelectric polarization. In AlGaN/GaN heterojunction, the charges induced due to the polarization can be evaluated [32]. The evaluated polarization charge and 2DEG density values with different Al compositions [33].

For AlGaN/GaN-based transistor devices, the source, drain and gate configuration are similar to standard MOS structure. Furthermore, the AlGaN/GaN-based devices can be integrated in the same way as complimentary metal-oxide semiconductor (CMOS). These devices are especially fabricated and designed for MMIC [34] and different power conversion applications [28]. For the designing of high power-gain cutoff frequency maximum oscillation frequency (f_{max}), the gate length should be approximately around 100 nm [7]. However, for power applications, the gate length (2 μm) and the gate-drain separation (7.5 μm) are higher in comparison to convention MOS for GaN-based HEMT. These higher lateral distances of GaN power HEMTs for power applications can increase the higher breakdown voltage, which is extremely useful in these applications. The switching frequency of device is associated with its structural design. However, the maximum operating frequency is also affected when devices are connected with circuits [33]. The switching performance of GaN devices at higher frequency can be highly affected by parasitic resistance and inductances in the interconnections.

4 GaN Base Station PAs

For higher bit rate communication, the most critical issues are the efficiency of base station broadband amplifier. The research and development of power amplifiers (GaN) are going on for nearly 25 years [36, 37]. The first commercial GaN HEMT technology has been launched in the market in 2005. The GaN HEMT devices with Doherty configuration feature high efficiency and sufficient reliability for the 5G base station PAs.

In MIMO systems, each antenna is having its separate PA. Therefore, for minimizing variation across cells, it is necessary to match the requirements of power and linearity. Advancement in base station PA (small-cell) of 5G GaN-based devices is needed. It will improve the compactness, weight, and reduce the cost with retaining high power and efficiency. It is very important to analyze the effect of self-heating, trapping, breakdown voltage, field plate design, and transconductance shape of GaN upon efficiency (PAE), power, operating frequency and other important parameters. The performance of GaN power transistor will be better than the current generation transistor like GaAs FET and Si-based LDMOS. However, this will need lot of effort in the design and analytical analyses of these high-power devices utilizing RF and DC techniques [30].

The Ids versus Vds characteristics and maximum gain of device of a 0.20 μm GaN 8×100 μm transistor have been reported in the literature [5]. It describes the design suitability of 5G PA GaN-based technologies. It has been generally observed that at 28 GHz frequency, parameter S21 is 5.372 dB and available gain is 14.091 dB. It can be observed from these figures available in the literature that GaN-based PA is having satisfactory gain at these operating frequencies [5].

Plenty of research has been done on GaN for 5G PA [31]. Primarily, GaN PA techniques have been established for satellite communications for different communication band. However, Ka-band is very suitable for 5G mm-Wave technology. It has been predicted by the researcher that GaN HEMT-based PAs will be commercially available as the draft of 1st 5G standards are made [38]. These GaN devices work around 28 GHz deliver operational power (several watts). Currently, GaN HEMTs are not available for commercial application at operating frequencies like 37 GHz and 39 GHz. However, some group has performed research at operating frequency 32 GHz [39]. This recognizes a key research gap for the development of GaN-based HEMT transistor.

The circuit layout of a device can be easily drawn [5]. The simulated analyses of these devices are performed by different tools. From the design perspective, it specifies that current GaN-based technology can be utilized in 5G (first-generation systems) [5].

5 GaN Frequency Synthesis

The 5G MIMO systems have a lot of antenna elements and increasing continuously. Currently, 5G MIMO used 32 or more antennas [40]. It is becoming difficult for generating and distributing coherent local oscillator (LO) power because the number of transmitters is increasing continuously. One of the ways to resolve these problems is to utilize PA for LO power amplification.

As we know, initially 5G signals started in the sub 6 GHz range. However, in current scenario, the lower frequency communication bands varying from GSM850/900 to DCS/PCS to LTE frequencies. So, the compatibility with lower frequency will be needed for better performance. Consequently, the issue of signal distribution in MIMO system and compatibility issue can be resolved. It can be resolved by utilizing frequency multiplication to deliver satisfactory power at the required frequencies. The above-discussed techniques can be useful for producing mm-wave 5G LO signal. High power GaN frequency multipliers can generate high frequency, high power LO signal. The generated output can be accurately allocated to every MIMO system utilizing a passive network. The GaN-based devices lead to nonlinear techniques [41]. The frequency multiplication technique allows GaN devices can deliver power at above current gain cut-off frequency (f_T) utilizing harmonic enhancement techniques. The growth of GaN-based technology for high-end applications without losing originality is one of the possible areas of technology growth.

6 Conclusion

This chapter analyses the GaN-based FET used in RF PAs. The discussion is focused on base station PAs made using GaN HEMTs in comparison to other technology. The leading technology for the base station is silicon-based LDMOSFET transistor. However, in current scenario LDMOSFETs, GaAs MESFETs, and GaN HEMTs are having equal share of one-third of the market. Though, as wireless market share is increasing, the market share of LDMOSFETs is going to shrink because 5G base stations work above 3.5 GHz where the LDMOS technology is not going to be a feasible alternate. It is very evident from the current scenario that GaN-based HEMT technology is going to replace the GaAs-based technology for 5G communication and wireless communication operating at mm-wave frequencies.

Author Contributions
This book chapter is written through contributions of all authors. Brajendra Singh Sengar and Amitesh Kumar have worked together on this publication and contributed equally. All authors have given approval to the final version of the manuscript. The authors declare no competing financial interest.

References

1. Yuk K, Branner G R and Cui C (2017) Future directions for GaN in 5G and satellite communications. In: Midwest symposium on circuits and systems, vol 2017, August, pp 803–806
2. Seidel A, Wagner J and Ellinger F (2020) 3.6 GHz asymmetric Doherty PA MMIC in 250 nm GaN for 5G applications. In: GeMIC 2020—Proceedings of the 2020 German microwave conference, vol 1, pp 1–4
3. Bhardwaj S, Kitchen J (2019) Broadband parallel Doherty power amplifier in GaN for 5G applications. In: 2019 IEEE IEEE Topical conference on RF/microwave power amplifiers for radio and wireless applications, PAWR 2019, pp 5–7
4. Del Gaudio A, Limiti E (2019) A 28 GHz MMIC Doherty power amplifier in GaN on Si Tech for 5G, pp 611–613
5. Yuk K, Branner GR, Cui C (2017) Future directions for GaN in 5G and satellite communications. In: Midwest symposium on circuits and systems vol 2017, August, pp 803–806
6. Gueguen P (2015) How power electronics will reshape to meet the 21st century challenges? In: Proceedings of the international symposium on power semiconductor devices and ICs vol 2015, June, pp 17–20
7. Sun R, Lai J, Chen W, Zhang B (2020) GaN power integration for high frequency and high efficiency power applications: a review. IEEE Access 8:15529–15542
8. Baliga BJ (1991) An overview of smart power technology. IEEE Trans Electron Devices 38:1568–1575
9. Letavic T, Arnold E, Simpson M, Aquino R, Bhimnathwala H, Egloff R, Emmerik A, Wong S, Mukherjee S (1997) High performance 600 V smart power technology based on thin layer silicon-on-insulator. In: IEEE International symposium on power semiconductor devices & ICs (ISPSD), pp 49–52
10. Chow TP, Omura I, Higashiwaki M, Kawarada H, Pala V (2017) Smart power devices and ICs using GaAs and wide and extreme bandgap semiconductors. IEEE Trans Electron Devices 64:856–873
11. Wong KY, Chen W, Liu X, Zhou C, Chen KJ (2010) GaN smart power IC technology. Phys Status Solidi Basic Res 247:1732–1734
12. Wu UKM , PP , Y-F AlGaNGa NHEMTs—An over view of device operation and applications.pdf
13. Klein K, Hoene E, Reiner R, Quay R (2019) Study on packaging and driver integration with GaN switches for fast switching. In: CIPS 2016—9th International conference on integrated power electronics systems
14. Pengelly RS, Turner JA (1976) Monolithic broadband gaas f.e.t. amplifiers. Electron Lett 12:251–252
15. Otsuka N, Kawai Y, Nagai S (2017) Recent progress in GaN devices for power and integrated circuit. In: Proceedings of International Conference on ASIC, vol 2017, October, pp 928–931

16. Pengelly RS, Wood SM, Milligan JW, Sheppard ST, Pribble WL (2012) A review of GaN on SiC high electron-mobility power transistors and MMICs. IEEE Trans Microw Theory Technol 60:1764–1783
17. Micovic M, Brown DF, Regan D, Wong J, Tang Y, Herrault F, Santos D, Burnham SD, Tai J, Prophet E, Khalaf I, McGuire C, Bracamontes H, Fung H, Schmitz A (2019) High frequency GaN HEMTs for RF MMIC applications
18. Hassan A, Ali M, Trigui A, Savaria Y, Sawan MA (2019) GaN-based wireless monitoring system for high-temperature applications. mdpi.com
19. Liu H, Zhu X, Boon C, Yi X, …, M M-I (2013) Design of ultra-low phase noise and high power integrated oscillator in GaN-on-SiC HEMT technology
20. Kuwabara T, Tawa N, … Y T-2017 I C (2017) A 28 GHz 480 elements digital AAS using GaN HEMT amplifiers with 68 dBm EIRP for 5G long-range base station applications. ieeexplore.ieee.org
21. Weber R, Schwantuschke D, … P B-2014 I M-S (2014) A 92 GHz GaN HEMT voltage-controlled oscillator MMIC. ieeexplore.ieee.org
22. Micovic M, Brown D, Regan D, … J W-2016 I (2016) High frequency GaN HEMTs for RF MMIC applications. ieeexplore.ieee.org
23. Zhang Y, Rodríguez M, Maksimović D (2016) Very high frequency PWM buck converters using monolithic GaN half-bridge power stages with integrated gate drivers. IEEE Trans Power Electron. 31:7926–7942
24. Garcia JA, Nieves Ruiz M, Vegas D, Pampin M, Mediavilla A (2017) UHF power conversion with GaN HEMT class-E2 topologies. In: 2017 IEEE compound semiconductor integrated circuit symposium, CSICS 2017 vol 2017-January, pp 1–4
25. Garcia JA, Marante R, De Las Nieves Ruiz Lavin M (2012) GaN HEMT class E2 resonant topologies for UHF DC/DC power conversion. IEEE Trans Microw Theory Technol 60:4220–4229
26. Sun R, Liang YC, Yeo YC, Wang YH, Zhao C (2017) Design of power integrated circuits in full AlGaN/GaN MIS-HEMT configuration for power conversion. Phys Status Solidi Appl Mater, Sci, p 214
27. Saito W, Takada Y, Kuraguchi M, Tsuda K, Omura I (2006) Recessed-gate structure approach toward normally off high-voltage AlGaN/GaN HEMT for power electronics applications. IEEE Trans Electron Devices 53:356–362
28. Cai Y, Zhou Y, Lau KM, Chen KJ (2006) Control of threshold voltage of AlGaN/GaN HEMTs by fluoride-based plasma treatment: From depletion mode to enhancement mode. IEEE Trans Electron Devices 53:2207–2214
29. Gozalvez J (2015) Samsung electronics sets 5G speed record at 7.5 Gb/s. IEEE Veh Technol Mag 10:12–16
30. Yuk KS, Branner GR, McQuate DJ (2009) A wideband multiharmonic empirical large-signal model for high-power GaN HEMTs with self-heating and charge-trapping effects. IEEE Trans Microw Theory Technol 57:3322–3332
31. Noh Y, Choi Y H, Yom I (2016) Ka-band GaN power amplifier MMIC chipset for satellite and 5G cellular communications In: Proceedings of the 2015 IEEE 4th Asia-Pacific conference on antennas and propagation, APCAP 2015 pp 453–456
32. Ambacher O, Smart J, Shealy JR, Weimann NG, Chu K, Murphy M, Schaff WJ, Eastman LF, Dimitrov R, Wittmer L, Stutzmann M, Rieger W, Hilsenbeck J (1999) Two-dimensional electron gases induced by spontaneous and piezoelectric polarization charges in N- And Ga-face AIGaN/GaN heterostructures. J Appl Phys 85:3222–3233
33. Xin Y, Chen W, Sun R, Shi Y, Liu C, Xia Y, Wang F, Li M, Li J, Zhou Q, Deng X, Chen T, Li Z, Zhang B (2019) Analytical switching loss model for GaN-based control switch and synchronous rectifier in low-voltage buck converters. IEEE J Emerg Sel Top Power Electron 7:1485–1495
34. Shaobing W, Jianfeng G, Weibo W, Junyun Z (2016) W-band MMIC PA with ultrahigh power density in 100-nm AlGaN/GaN technology. IEEE Trans Electron Devices 63:3882–3886

35. Shaobing W, Jianfeng G, … W W-I T (2016) W-band MMIC PA with ultrahigh power density in 100-nm AlGaN/GaN technology. ieeexplore.ieee.org
36. Anon GaN on SiC Foundry Services| Wolfspeed
37. Anon Global Communication Semiconductors, LLC
38. Anon Microelectronics Products and Services—Northrop Grumman
39. Micovic M, Brown D F, Regan D, Wong J, Tang Y, Herrault F, Santos D, Burnham S D, Tai J, Prophet E, Khalaf I, McGuire C, Bracamontes H, Fung H, Schmitz A (2019) High frequency GaN HEMTs for RF MMIC applications
40. Sadhu B, Tousi Y, Hallin J, Sahl S, Reynolds SK, Renstrom O, Sjogren K, Haapalahti O, Mazor N, Bokinge B, Weibull G, Bengtsson H, Carlinger A, Westesson E, Thillberg JE, Rexberg L, Yeck M, Gu X, Ferriss M, Liu D, Friedman D, Valdes-Garcia A (2017) A 28-GHz 32-element TRX phased-array IC with concurrent dual-polarized operation and orthogonal phase and gain control for 5G communications. IEEE J Solid-State Circuits 52:3373–3391
41. Yuk KS, Branner GR, Wong C (2010) High power, high conversion gain frequency doublers using SiC MESFETs and AlGaN/GaN HEMTs, pp 1–1

The manufacturer's authorised representative in the EU is Springer Nature Customer Service Centre GmbH, Europaplatz 3, 69115 Heidelberg, Germany. If you have any concerns regarding our products, please contact ProductSafety@springernature.com

Printed and bound by CPI Group (UK) Ltd, Croydon, CR0 4YY
30/09/2026
02225264-0013